PAKISTAN'S LOW-CARBON ENERGY OUTLOOK AND TECHNOLOGY ROAD MAP

NOVEMBER 2024

ASIAN DEVELOPMENT BANK

6 ADB Avenue, Mandaluyong City, 1550 Metro Manila, Philippines
Tel +63 2 8632 4444; Fax +63 2 8636 2444
www.adb.org

ISBN 978-92-9270-968-6 (print); 978-92-9270-969-3 (PDF); 978-92-9270-970-9 (ebook)
Publication Stock No. TCS240489-2
DOI: http://dx.doi.org/10.22617/TCS240489-2

Notes:
In this publication, "$" refers to United States dollars, "¥" refers to Japanese yen, and "PRs" refers to Pakistan rupees.
ADB recognizes "China" as the People's Republic of China.

Cover design by Mike Cortes.

CONTENTS

TABLES AND FIGURES

Tables

Figures

ACKNOWLEDGMENTS

The *Pakistan Low-Carbon Energy Outlook and Technology Road Map* is an in-depth analysis commissioned by the Asian Development Bank (ADB) and developed by the Energy Sector Office, Sectors Group. This report outlines a comprehensive long-term pathway for the country's energy sector and associated greenhouse gas emissions, focusing on low-carbon energy technologies and efficiency measures. It employs a rigorous modeling approach utilizing the MESSAGE-ix energy model developed by the International Institute for Applied Systems Analysis.

The study was carried out under the leadership of Kee-Yung Nam, principal energy economist, with supervision from Priyantha Wijayatunga, senior director of the Energy Sector Office, Sectors Group. The report was prepared, and the modelling was conducted by the management and staff of PricewaterhouseCoopers International Limited. David Morgado, senior energy specialist, coordinated and reviewed the content. John Weiss, serving as the technical editor, synthesized the two documents into an integrated report.

A team of both international and national experts offered significant insights as peer reviewers. Charity L. Torregosa, senior energy officer, managed the production process. Maria Dona Aliboso, associate project officer and Angelica Apilado, associate project analyst provided administrative support. Ma. Theresa Mercado copyedited the report, while Jonathan Yamongan did the layout. Jess Alfonso Macasaet, proofreader and Lawrence Casiraya, page proof checker also contributed to the production of this publication. The study also gained valuable insights and feedback from colleagues at the ADB Energy Sector Office, Sectors Group.

This study is supported by funding from the Clean Energy Fund under the Clean Energy Financing Partnership Facility, the People's Republic of China Poverty Reduction and Regional Cooperation Fund, the Urban Climate Change Resilience Fund, and the E-Asia Knowledge Partnership Fund.

ABBREVIATIONS

ADB	Asian Development Bank
AEDB	Alternative Energy Development Board
BAU	business-as-usual
CAGR	compound annual growth rate
ESCO	energy service company
GDP	gross domestic product
GHG	greenhouse gas
IEA	International Energy Agency
IFC	International Finance Corporation
IIASA	International Institute for Applied Systems Analysis
JICA	Japan International Cooperation Agency
LNG	liquefied natural gas
NDC	nationally determined contribution
NEECA	National Energy Efficiency and Conservation Authority
PES	primary energy supply
PV	photovoltaic
RLNG	regasified liquefied natural gas
TFED	total final energy demand
TPES	total primary energy supply
TRL	technology readiness level

Weights and Measures

GW	gigawatt
$MtCO_2e$	million tons of carbon dioxide equivalent
Mtoe	million tons of oil equivalent
MW	megawatt
TWh	terawatt-hour

EXECUTIVE SUMMARY

Pakistan is the fifth most populous country in the world with a population of around 236 million and a per capita gross domestic product (GDP) of about $1,597, based on World Bank estimates for 2022. The country's GDP grew to 5.8% in 2021 and 6.1 % in 2022, indicating a remarkable recovery from the coronavirus disease pandemic. However, the country is still challenged by domestic and international events like the 2022 floods and the Russian invasion of Ukraine. The International Monetary Fund expects Pakistan's economy to perform better in the current and coming fiscal years.

The government is making concerted efforts to revive its ambition of achieving an upper middle-income country status by 2047. As a step in this direction, in July 2023, the Government of Pakistan approved the Pakistan Investment Policy 2023 to enhance both foreign and domestic investments and to achieve sustainable and inclusive economic growth, industry innovation, and infrastructure development. The policy focuses on simplifying business regulations, addressing investment disputes, and granting incentives on establishing businesses in the country.

Such economic growth, accompanied by Pakistan's large and growing population and increasing urbanization will result in increase in energy demand, which if not met will become a major constraint to growth. Currently a significant share of primary energy supply comes from fossil-fuel-based sources, over 60% in 2019, with coal use growing rapidly. While total greenhouse gas (GHG) emissions are low on a per capita basis, the energy sector has been an important source of emissions, which have roughly doubled since 2000. It currently contributes nearly half of total national emissions. A key policy challenge is therefore to maintain an acceptable rate of economic growth, meeting growing energy demand adequately and in an affordable manner, while minimizing the damage caused by emissions and meeting government targets and its international commitments under the Paris Agreement on Climate Change.

Pakistan is highly vulnerable to the adverse effects of climate change and is among the top 10 countries most affected by climate change. The government is committed to addressing the problem of rising emissions. In 2016 as part of the climate change negotiations under the nationally determined contributions, the government proposed to reduce up to 20% of national GHG emissions by 2030, taking 2015 as the base year. In 2021, the government agreed to revised NDCs that extended this target to an overall 50% reduction by 2030, where 15% is unconditional and the remaining 35% is conditional on receipt of adequate international financial assistance.

To meet this commitment, the energy sector will need to evolve on a different path. This study aims to assist this process by setting out a detailed pathway up to 2050 in terms of technologies and efficiency measures. It does this a using rigorous modeling approach by applying the MESSAGE-ix energy model developed by the International Institute for Applied Systems Analysis. This model, which is widely used in energy sector planning, is a cost-minimization exercise, subject to defined constraints. The benefits from pursuing a green energy policy are illustrated by running the model for two scenarios. One, a business-as-usual (BAU) scenario, assumes the continuation of current and historic policy and technology trends, which include a gradual shift toward renewable energy sources. The other is a low-carbon scenario, which assumes significant and rapid deployment of emerging low-carbon technologies and options to reduce GHG emissions. Both scenarios employ the same projections on population and per capita GDP growth. The difference between the two scenarios in terms of emissions provides a measure of the climate change benefits from a low-carbon energy strategy.

Details of the technologies, efficiency measures, and the investment costs associated with the low-carbon pathway are discussed. The study also tabulates various barriers and enablers for the application of the various technologies that are part of the cost-minimization pathway. It also identifies the investment needs for each technology, along with potential modalities for sourcing the necessary financial resources for their deployment. Following are the characteristics of the low-carbon scenario:

Higher Investment but Lower Emissions

The low-carbon scenario would require an additional investment of $191.8 billion (in 2022 prices) between 2020 and 2050 over the BAU scenario, so the investment commitment is substantial. Energy sector GHG emissions rise in both scenarios but under the low-carbon scenario total energy sector GHG emissions would be almost 23% lower in 2030 and 36% lower in 2050, as compared with the BAU scenario. Hence the same growth rate is possible, but with around one-third less emissions.

Lower Energy Demand due to Greater Efficiency in User Sectors

Total final energy demand is estimated to be 12% lower by 2050 under the low-carbon scenario. Energy efficiency improvement has a critical role in reducing demand. From the supply side, a switch toward more efficient power production can also contribute to the reduction in energy demand. Out of the total energy demand reduction of about 51 million tons of oil equivalent (Mtoe) in 2050, the main reductions are in industry (47% of the total) and transport (39%). The low-carbon scenario shows a strong effect of electrification and fuel-switching. The majority of the emissions mitigation in the low-carbon scenario relative to that in the BAU scenario (91% in 2030 and 75% in 2050) comes from demand-side initiatives by users. The key mitigation opportunities from the demandside lie in the coal-to-gas transition in industry, electrification of transport, and electricity replacing use of gas for cooking.

Fuel-Switching to Cleaner Alternatives

A key part of the low-carbon scenario involves a significant shift to cleaner fuels. The key areas with high potential for emissions mitigation are nuclear-based power generation, hydropower generation, and solar photovoltaic and wind power generation. The role of nuclear power, renewables, and natural gas is crucial to support the clean energy transition. Under the low-carbon scenario in 2050, around 26% of the total primary energy supply (TPES) would come from natural gas and about 22% collectively from nuclear fuels, solar, hydro, and wind. Coal would not be phased out immediately, although new investment in coal plants would stop in 2030 and would take 23% of TPES in 2050. This is in contrast with the BAU scenario where coal would take 43% of TPES in 2050, gas 16%, and nuclear fuel, hydro, solar, and wind combined 12%.

Electrification and Change Within the Electricity Sector

Greater use of electricity in transport, the residential sector (including cooking), and agriculture is a key source of energy efficiency improvements. This is illustrated in the share of electricity in final energy consumption between the BAU and low-carbon scenarios. In the BAU scenario, this is 11% in 2050, but with a higher level of electrification in the low-carbon scenario, this increases to 16%. For electrification to be a successful strategy in meeting decarbonization targets, it is essential to move from fossil-based power generation to renewable energy-based generation. The low-carbon scenario implies a phased transition from coal to renewables. Under the BAU scenario, the country continues to invest in power generation capacity based on natural gas and coal, with moderate capacity additions in renewable-based generation. In contrast, under the low-carbon scenario, the power sector undergoes a major restructuring with the ending of investment in coal and gas-based plants from 2030 and a significant push toward nuclear, solar PV, onshore wind, and hydro-based power generation. By 2050, 81% of electricity capacity comes from renewables (hydro, solar, and wind) with only 3% from coal-based plants. In terms electricity generation by 2050 over 90% of supply comes from clean energy (61% from renewables and 29% from nuclear) with coal and biomass accounting for the majority of the remainder. This contrasts with the structure of the power sector in the BAU scenario, where in 2050, coal-based plants still account for 17% of capacity and 26% of electricity generated.

Import Dependence Continues but Composition Changes

Although the low-carbon scenario results in lower energy requirements than the BAU scenario, import dependency (imports in total primary energy supply) in 2050 would be 61% as compared to 53% in the BAU scenario. In absolute terms, the low-carbon scenario would require an almost similar quantum of energy imports, but the imported fuel composition would be different. In the BAU scenario, by 2050, oil would account for 40% of total energy imports followed by gas (29%) and coal (14%). Under the low-carbon scenario, the highest growth in energy imports would be in natural gas and by 2050, natural gas would account for about 40% of total energy imports, followed by oil (32%) and uranium (20%). Energy security planning in the country would need to account for this level of import dependence in primary energy supply, with its associated geopolitical risks to energy security.

Key Technologies, 2025–2050

The key technologies along the low-carbon pathway—in terms of their addition to capacity—will be hydro, nuclear, solar, and wind. For nuclear, there are issues relating to high capital investment and operating costs and safety concerns, coupled with the absence of a clear legal basis for managing civil nuclear liability, especially in the context of an accident with transboundary consequences. The government needs to prioritize using reactors with high safety levels and simplified systems. However, despite the long lead time and high capital cost involved, if managed properly, the technology offers a way of expanding clean generation capacity on a large scale.

There is vast untapped potential for hydropower in the country. Although it is a mature technology, investment cost is high and there are challenges in implementation due to the disruption to communities and ecosystems caused by large dam projects. Due to the long life of hydro dams, unit electricity costs can be relatively low and hydro supply offers the potential to offset the variability that will be introduced into the grid by a rising share of renewables in energy supply.

For wind and solar energy there is considerable scope for expanding capacity and, in the case of solar, taking advantage of the country's levels of sunlight. The key issues are aligning financial incentives for private sector investors with the economic benefits from renewable energy production, so that investment in renewables is encouraged, and developing manufacturing capacity in the country for at least some of the components required. Currently there is relatively little such capacity and where local components are produced, they are high cost.

High Investment Costs Will Need Major External Funding

These expansion plans imply major investment commitments. In the low-carbon scenario between 2020 and 2050, hydropower generation requires an investment of $153 billion, nuclear power $103 billion, wind power $62 billion, and solar $51 billion. To handle growing power demand and maintain grid stability, investment in transmission and distribution of $22 billion is also required. These large investments in the power sector are in addition to the investment required in the transport and domestic sectors to achieve energy efficiency savings. Achieving such an ambitious investment program will be challenging. Direct government funding is likely to be limited, so a majority of funds will have to come from the foreign and domestic private sector—through direct investment, equity funding, bank credits and bond issues—and international financial assistance. The investment climate must be sufficiently encouraging for this large-scale funding to be possible.

Policy Reform

This will require major policy initiatives to strengthen the system of financial intermediation, align incentives for the private sector with the goals of environmental policy, ensure the regulatory system supports these goals effectively, and develop a system of project preparation and implementation that meets the requirements of multilateral and bilateral donor organizations. Only if there is major progress in these areas will it be realistic to think that the low-carbon scenario outlined in the study can be fully funded and implemented. In addition, from the side of energy delivery, effective system planning, adequate mechanisms and incentives for supply- and demand-side flexibility, short-term system balancing, and stability procedures will be required to support a transition to a low-carbon pathway.

INTRODUCTION

The energy resources of Pakistan consist of fossil fuels such as coal, oil and gas, nuclear fuel, and renewable energy such as hydropower, solar, wind and biomass. The hydropower potential of the country is large and is estimated to be about 50–60 gigawatts (GW).[1] Pakistan currently imports one-third of its primary energy sources including coal, oil, and natural gas (ADB 2019). Continued dependence on energy imports has raised challenges related to high energy costs, while about 40 million people (20% of the population) do not have access to electricity (IEA, Pakistan 2021). To curb import dependence and rising greenhouse gas (GHG) emissions, decentralized solar energy systems, combined with investment in other green sources of energy, are essential (ADB 2019).

Over the years, the country had diversified its energy supply, but by 2019, more than 62% of primary energy was supplied from fossil-based energy sources, with coal having the highest growth (IEA, Pakistan 2021). Emissions from energy-related sources have almost doubled since 2000. This growth was driven by the previous energy policy, which relied on coal imports and increasing domestic coal production to meet the growing energy demand. If expansion plans are based on coal-based capacity, emissions will continue to rise rapidly. Between 2020 and 2050, the economy is estimated to grow by more than three times and the population is expected to increase by 85 million. Rapid economic growth and rising levels of urbanization will lead to higher energy requirements in future and the key challenge will be to supply this energy with a sustainable level of emissions.

According to the Global Climate Risk Index, Pakistan is among the top 10 climate-affected countries in the world. The National Disaster Management Authority of Pakistan estimated that over 1994–2013, an average annual economic loss of $4 billion could be attributed to climate adversities (UNFCCC 2015). The Government of Pakistan recognizes the adverse impacts of climate change and is committed to addressing the problem of rising emissions. In 2016, as part of the climate change negotiations, under the nationally determined contributions (NDCs) the government proposed to reduce up to 20% of national GHG emissions by 2030, taking 2015 as the base year. In 2021, the government agreed to revised NDCs that extended this target to an overall 50% reduction by 2030, where 15% is unconditional and the remaining 35% is conditional on receipt of adequate international financial assistance. (UNFCCC 2021)

This report explores future energy and emissions pathways for Pakistan's energy sector up to 2050. It is divided into four sections. Chapter 1 sets out the model used to project two alternative scenarios for the evolution of energy supply, demand, and emissions. One is based on an assessment of the continuance of current trends (the business-as-usual [BAU] scenario) and the other takes account of new policy targets and gives greater emphasis to green technologies and energy efficiency (the low-carbon scenario). Chapter 2 gives the results of the model. These results are not precise predictions of what will happen, but rather outcomes that would occur if the assumptions underlying the scenarios and the coefficients and relationships underlying the model match reality. Chapter 3 gives the different technologies and efficiency measures that are required to meet the future low-carbon pathway. Chapter 4 gives estimates of cost of investment requirements if the low-carbon pathway is to be achieved and discusses the sources of funding for this future investment. Chapter 5 gives the conclusions and the implications of a low-carbon pattern of development in Pakistan.

CHAPTER 1

THE MODEL AND TECHNOLOGIES

The model used for this study is the MESSAGEix modeling framework developed by the International Institute for Applied Systems Analysis (IIASA). MESSAGEix is a versatile, open source, dynamic systems optimization model representing an energy system that converts energy for end-use demand in the form of heat, light, or kinetic energy. The model allows for interdependencies between resource endowments, extraction rates, imports and exports, the generation of electricity and the conversion of fuels, transportation, transmission, and distribution. As an optimization model it determines the least-cost solution subject to constraints over predefined fixed time periods.

The MESSAGEix model is used widely across countries for national and subnational energy transition assessments. It has the flexibility to evaluate and assess mitigation potential across sectors for various low-carbon development scenarios and allows an assessment of the impacts of alternative scenarios across the national economy and various sectors. The five sectors considered here are agriculture, commercial, industry, residential, and transport. To map the future pattern of the energy transition, two scenarios are developed: one reflecting current trends (BAU scenario), which includes some renewables, and the other a low-carbon path (low-carbon scenario), with a greater emphasis on renewables and energy efficiency. Both scenarios are be based on the same gross domestic product (GDP) and population forecasts, with the difference being the strength of policy commitment to a clean energy transition.

The model uses population and GDP projections derived from the Shared Socioeconomic Pathways used frequently in analyses that estimate global emissions pathways. The analysis in this study uses the mid-range Shared Socioeconomic Pathways 2 case as the basis for both scenarios (IIASA 2018).

Table 1 and Table 2 summarize the data from the Shared Socioeconomic Pathways 2 scenario for Pakistan. Population is expected to grow at an annual rate of 1.2% and GDP at 3.9% over the modeling period 2020 to 2050.

Table 1: Population Projection for Pakistan
(million)

2020	2025	2030	2035	2040	2045	2050
207.34	224.14	239.68	254.2	267.97	280.89	292.57

Source: Riahi, K. et al. 2017. The Shared Socioeconomic Pathways and their energy, land use, and greenhouse gas emissions implications: An overview. *Global Environmental Change*.

Table 2: Gross Domestic Product Projection for Pakistan
($ billion 2010/year)

2020	2025	2030	2035	2040	2045	2050
640	749	892	1,079	1,319	1,620	1,988

Source: Riahi, K. et al. 2017. The Shared Socioeconomic Pathways and their energy, land use, and greenhouse gas emissions implications: An overview. *Global Environmental Change*.

Business-as-usual scenario. This assumes the continuation of recent trends in technological structural change and resource consumption of energy commodities such as coal, oil, gas, biomass, hydro, and emerging renewables; and cleaner technologies such as solar, wind, and nuclear. This means there is only a gradual shift toward renewables and cleaner energy.

Low-carbon scenario. This assesses options for achieving energy and emissions reduction trajectories based on an enhanced ambition for low-carbon development in support of the Paris Agreement. New policies and targets linked with these are incorporated in the scenario. Measures to achieve these include demand-side electrification, supply- and demand-side efficiency improvements, and fuel-switching across supply and demand sectors toward carbon-neutral or low-carbon fuels such as natural gas, biomass, and off-grid renewable energy generation. The NDC targets are not explicitly provided as an input, rather the scenario outcomes are evaluated to see how close they come to meeting the NDC targets. The list of low-carbon targets used to define the low-carbon scenario is given in Table 3.

The modeling time frame considered is from 2020 to 2050, with data given at 5-year intervals. These scenarios are not purely cost-optimization scenarios, as assumptions about policies and technologies in use are imposed as constraints. In the case of the BAU scenario, this is to maintain consistency with current and historic trends and in the low-carbon scenario, to maintain consistency with the low-carbon policy initiatives and associated targets. This means that in the low-carbon scenario costs are minimized, subject to these imposed constraints in terms of targets as specified in Table 3.

Table 3: Summary of Low-Carbon Interventions Adopted in the Low-Carbon Scenario

Scenario	Target description
Nationally Determined Contribution	Reduce emissions by 20% compared to the baseline projections by 2030*
Fuel shift	More than 45% share of gas in the industry thermal energy demand by 2030 and more than 50% by 2050
	About 20% of road and rail-based passenger transportation energy demand met by electric vehicles by 2050
	10% share of solar-based pumping systems within the agriculture pumping-based final energy demand in 2030 and close to 25% by 2050.
Renewable Energy	40% non-fossil penetration in the power plant capacity mix by 2030 and 60% including hydro
	5 gigawatts (GW) of installed wind power capacity by 2030
	10 GW of cumulative installed solar photovoltaic capacity (on-grid) by 2030
	About 30% of residential electricity demand met from off-grid solar by 2050
Energy Efficiency	Annual 0.1% improvement in vehicle fuel efficiency annually until 2050; Annual 0.2% improvement in industrial energy efficiency till 2050

* Pakistan updated its nationally determined contribution target in 2021 to commit a voluntary contribution of overall 50% reduction of Pakistan's projected emissions by 15% unconditional and 35% conditional.

Source: Compiled by authors.

After the declaration of the National Energy Efficiency and Conservation Authority (NEECA) Act in 2016, Pakistan has recently passed the National Energy Efficiency and Conservation Policy 2023 that identifies and mandates enforcement of several energy efficiency measures across various sectors, including ensuring compliance with minimum energy performance standards and labeling for appliances, mandating energy conservation building codes, and setting energy consumption thresholds for designated consumers in industries.

Assumptions on improvements in energy-efficient technologies, aligned with current policies and trends, are common across BAU and low-carbon scenarios. Additional energy efficiency improvement under the low-carbon scenario result from other strategies like electrification and fuel switching.

Technologies Incorporated in the Model

The model developed for the analysis considered a range of technologies, including existing and upcoming technologies. As part of the low-carbon scenario, the model investigates potential technologies which can play a key role in decarbonizing the economy in a cost-effective manner, considering the country's international and national commitments and resource endowments. Based on a literature review and stakeholder consultations, technologies relevant to the country's context were identified and tested in the model. The key technologies are summarized in Table 4, with reference given to their links with current policy. The Appendix discusses hydrogen-based developments, with a potential for future use, but not included in the low-carbon scenario as their commercial readiness is not yet adequate and there is no clarity on plans for development of green hydrogen in the country. In addition, the country is prioritizing economic growth over investing in expensive hydrogen technologies, which may be a challenging task for the country in the current macroeconomic situation.

Table 4: Key Technologies Considered in the Model

Technology	Description	Country Context	Policy Link
Coal Ultra-Supercritical (USC)	Clean coal technology requires less coal per megawatt-hour (MWh) and an operating efficiency of more than 45%, thereby additionally reducing operating costs.	The target is to add 8.4 gigawatts (GW) of pulverized coal combustion power plants by 2021 and another 15 GW by 2030. Pakistan's national climate change policy 2012 recommends retrofitting coal plants with carbon capture and storage (CCS) to reduce greenhouse gas (GHG) emissions. SEPCOIII in collaboration with General Electric (GE) power to build a 660-megawatt (MW) power plant in Deh Ghangiaro, which was expected to start commercial operations in 2021. The People's Republic of China (PRC) announced a few weeks ahead of the 2021 United Nations Climate Change Conference (COP26) that it will end investments in overseas coal projects, which could mean the cancellation of coal-fired pipeline projects.[a]	National Climate Change Policy 2012

continued on next page

Table 4 *continued*

Technology	Description	Country Context	Policy Link
Coal Supercritical Power Plant (PPL)	Clean coal technology requires less coal per MWh and operating efficiency of more than 42% compared to standard coal-fired plants that operate at around 33%.		National Climate Change Policy 2012
Coal Subcritical PPL	A conventional process where coal is processed under relatively lower pressure and temperature conditions. Costs are low; however, emissions are higher, and efficiency is significantly lower at around 33%.	The Government of the PRC has scrapped earlier plans to expand coal capacity as a part of investments abroad.[b] A 300-megawatt (MW) Gwadar plant is set to be built and funded by the PRC state-owned entities. Coal phase out in Pakistan may be slower than envisaged.	National Climate Change Policy 2012
Gas Combined Cycle Power Plant	The combined cycle (CC) process involves extracting heat from both the input and exhaust stream. This helps reduce energy wastage, reduce emissions, and increase efficiency to over 60% compared to single cycle efficiency of 34%.	Siemens and Harbin Electric collaborated with K-Electric to set up a 900 MW combined cycle power plant (CC PPL) in Karachi. Investments in gas CC plants are driven primarily from the PRC, however, in recent years, local investors such as private banks have started funding installations.	
Co-Gasification of Coal and Biomass	Emerging clean technology that can achieve high thermodynamic efficiency with relatively low carbon dioxide (CO_2) emission. The main advantage of co-gasification of coal and biomass is higher efficiencies of carbon conversion to syngas or electricity in small as well as large projects with economic value.	The total estimated biomass potential of Pakistan is 50,000 gigawatt-hours (GWh)/year. Pakistan has around 3,064 million tons of coal reserves. Also, it is the world's fifth largest producer of sugarcane. Further, the collective processing residues of major crops such as cotton stalks, wheat straw, rice straw, sugarcane trash, and corn stalk amount to around 25.27 million tons. (Pakistan, Policy for Development of Renewable Energy for Power Generation, 2006).	Policy for Development of Renewable Energy for Power Generation 2006 (for biomass gasification)
Energy Recovery from Liquefied Natural Gas (LNG) Regasification	The process through which discarded cold energy available during the LNG regasification process can be recovered and used as a useful cooling source. The recovered energy can be used as electricity, cooling (for cold chain/refrigerated trucks), and air conditioning.	Although Pakistan was a new market for LNG in 2015, the growth of LNG imports has been substantial. The LNG regasification capacity in the country is around 10 million tons per year (MTPA) with 54% regasification utilization. Pakistan has already commissioned two LNG receiving terminals of 3.8 MTPA capacity in 2015 and 5.7 MTPA in 2017. Further, LNG-fired power capacity is set to rise to 10.9 GW by 2030 from about 6.79 GW in 2020.	National Climate Change Policy 2012 (for waste heat recovery in general)

continued on next page

Table 4 *continued*

Technology	Description	Country Context	Policy Link
Coal Integrated Gasification Combined Cycle (IGCC)	Next-generation power generation system with enhanced efficiency and environmental performance due to the combination of coal gasification and gas turbine combined cycle.		
Carbon Capture and Storage	Technology addition that would assist in separating CO_2 from fossil fuel combustion process and subsequently collecting it and dumping it underground or at sea. It is an integrated concept that involves capture, transport, and storage components.	Pakistan is planning to add 8.4 GW of pulverized coal combustion power plants by 2021 and another 15 GW by 2030. The National Climate Change policy 2012 recommends retrofitting coal plants with CCS to reduce GHG emissions.	National Climate Change Policy 2012
Diesel Generator	Combination of diesel engine and electric generator to produce electricity. Typically, only used as a backup in places without grid connectivity as fuel efficiency is low.		
Pumped Hydro Storage (PHS)	PHS stores electrical energy as the potential energy of water. It involves pumping water into a large reservoir at a high elevation—usually located on the top of a mountain or hill	Pakistan has a hydro potential of around 50–60 GW, but the total installed hydro-capacity was around 10,752 MW as of June 2022. PHS is a feasible option to improve generation as well as the availability of water for irrigation.	Indicative Generation Capacity Expansion Plan (2022)
Geothermal Power Plant	Geothermal energy is a natural heat resource stored in the interior of the earth which requires heat transportation to the surface for utilization. Once transported, heat can be used to generate electricity.	Pakistan is estimated to have the potential of generating 100,000 MW at the cost of \$0.05–\$0.10 per unit. Hot springs and mud volcanoes exist within the seismic belt of Pakistan. However, Pakistan is yet to invest significantly to tap into these sites.	There is no policy or framework to drive geothermal power development
Nuclear Power Plant	A facility that converts atomic energy into usable power. The heat produced by nuclear fission in a reactor is used to drive a turbine to power electric generators.	Nuclear power is currently supplied at 3,620 MW. The PRC-funded Karachi Nuclear Power Plant, KANUPP-2 with the capacity of 1.1 GW, entered commercial operation in 2021. Two more plants with capacity of 1,100 MW each are currently in various stages of implementation. Recently, in 2023, the third unit of the KANUPP, which has 1.1 GW of power generating capacity has also been inaugurated.[c] A nuclear power generation capacity of 40,000 MW is included in Nuclear Energy Vision 2050. Eight more sites have been selected with four nuclear power plants each, to give a planned total of 32 nuclear power plants by 2050.	National Power Policy 2013; Indicative Generation Capacity Expansion Plan (2022); Nuclear Energy Vision 2050

continued on next page

Table 4 *continued*

Technology	Description	Country Context	Policy Link
Wind Power Plant	Wind is used to drive mechanical turbines to generate power. An offshore setup typically has greater power output and lesser ecological impact. However, it has higher investment and maintenance costs compared to onshore plants.	Studies have shown that Pakistan has a wind potential of 132 GW[d] with the Gharo-Jhimpir wind corridor identified as the region with maximum wind energy potential, estimated at 43 GW. Pakistan has 1,845 MW of installed capacity (2022) with the government pushing for greater adoption. The government is proposing tax exemption for manufacturers while also keeping power tariffs at a minimum and providing feed-in tariffs, the first of which was seen in Sapphire Wind Farm. The target is 6% of total power generation from wind energy by 2030.	Alternative and Renewable Energy Policy (2019); Indicative Generation Capacity Expansion Plan (2022)
Biomass Power Plant	Wood waste or other waste is burned to produce steam that runs a turbine to make electricity. With pollution controls and combustion engineering net emissions from biomass are lower than conventional fossil fuels.	The majority of biomass in Pakistan is sourced from agricultural waste residue. The bioenergy potential[e] from sugarcane and cotton waste alone is estimated to be over 12.5 terawatt-hours (TWh)/year. Currently, bagasse plants are the primary source of bioenergy with an installed capacity of over 900 MW. Municipal waste is also growing with an estimated potential of 13.5 TWh/year. Lahore waste management company approved the development of a 40 MW plant in 2018 to tap into the same. A joint venture was accepted by National Electric Power Regulatory Authority in 2021 to build the plant.	Alternative and Renewable Energy Policy (2019)
Solar Photovoltaic (PV) Grid	Photovoltaic panels or arrays are connected to the utility grid through a power inverter unit allowing them to operate in parallel with the electric utility grid. Extra power generated can be stored in batteries or can be fed back into the grid.	Despite having one of the highest solar potentials in the world, Pakistan has been relatively slow to invest in solar technology. Current installed solar PV capacity is 600 MW (NTDC 2022). Construction of Quaid-E-Azam Solar Park with a capacity of 1 GW commenced in 2013, however, due to political tensions and lack of support from the government, only 400 MW capacity has been installed with the remaining capacity installation halted.	Alternative and Renewable Energy Policy (2019); Indicative Generation Capacity Expansion Plan (2022); Net Metering Programme (2015)

continued on next page

Table 4 *continued*

Technology	Description	Country Context	Policy Link
Solar-Powered Irrigation Systems (Solar Irrigation Pumps)	Agricultural pumps use solar energy, electricity, or diesel as fuel for pumping water for irrigation.	Out of around 1.3 million tube wells and lift pumps in Pakistan, around 83% are diesel-powered irrigation and the rest (17%) are electric. The International Water Management Institute launched the Solar Irrigation for Agricultural Resilience project, in 2016, to replace diesel-powered irrigation pumps with solar irrigation systems throughout Bangladesh, India, Nepal, and Pakistan.[f]	
Concentrated Solar Power (CSP)	CSP systems use mirrors to concentrate the sun's energy to drive turbines or engines that can generate electricity. The thermal energy generated from a CSP plant can be stored and used to produce electricity as and when required (day or night).	CSP or integrating it with indigenous coal or biomass can bring a considerable boost to the overall power system capacity while reducing dependence on imported fuel and emissions by the transformation of nonrenewable technology into low renewable technology. Sites in Sindh and plain areas of Balochistan have been identified as potential areas for investment in CSP.	
Floating Solar	Floating solar PV is a solar PV application in which PV panels are designed and installed to float on water bodies such as reservoirs, hydroelectric dams, industrial ponds, water treatment ponds, mining ponds, lakes, and lagoons	Pakistan has several large water reservoirs such as Tarbela, Mangla, Ghazi Barotha, and others. Further, the total installed hydro-capacity by 2018 was around 7,477 MW. In addition, the government intends to add 18–20 GW of new hydro plants. These new hydro plants are being considered as potential sites for floating solar power plants.	National Climate Change Policy 2012 encourages the development of renewable energy sources including solar and hydropower
Decentralized Generation and Mini-Grids	A mini-grid is a system having a renewable energy (RE)-based electricity generator (with the capacity of 10 kilowatts [KW] and above), supplying electricity to a target set of consumers (residents for household usage, commercial, productive, industrial, and institutional setups etc.) through a public distribution network.	The geographic location of rural parts of Pakistan is such that rural communities are located at long distances from each other in remote locations. In such cases, mini and micro off-grid networks can be a viable alternative by tapping the renewable potential of solar, hydropower, and others.	National Power Policy, 2013 - Policy for Development of Renewable Energy for Power Generation; Pakistan Poverty Alleviation Fund initiative by Germany's KfW

continued on next page

Table 4 *continued*

Technology	Description	Country Context	Policy Link
Battery Energy Storage Systems (BESS)	Energy storage could apply to different technologies ranging from pumped hydro storage, flywheels, supercapacitors, compressed air, thermal energy storage, and batteries. Advanced energy storage technologies are capable of dispatching electricity within seconds and can provide power backup ranging from a few minutes to long hours.	The present standards and conditions of the electricity distribution in Pakistan are inadequate to meet the rapidly growing power demand, according to National Transmission and Dispatch Company, Pakistan. A reliable and continuous power supply requires energy storage systems such as BESS. Further, BESS facilitates stabilizing electricity frequency variations. In line with this, under the Asian Development Bank High-Level Technology Fund, the country is piloting the first large-scale grid-connected lithium-ion battery energy storage system.	
Electric Vehicles	Electricity-based technology that would become an alternative to internal combustion engine vehicles.	As per the National Electric Vehicle Policy (NEVP), Pakistan has planned to increase the number of electric vehicles (EVs) to 100,000 cars; 500,000 two and three-wheelers; and 1,000 buses and trucks by the year 2025. Further, it is planned that around 30% of the total vehicles in Pakistan would be EVs by 2030, and the share would be increased to 90% in 2040.	NEVP – The government has set goals, provided tax incentives, custom duty exemptions, and planned public investments
High-Efficiency Residential and Commercial Heating, Ventilation, and Air-Conditioning (HVAC)	Use of inverter technology and fixed speed air-conditioners for better performance of the air-conditioners along with a common metric and standards and labeling program to increase savings and quality	In 2018, the residential air conditioner demand was around 777,000 units and commercial air-conditioner demand was around 47,000 units in Pakistan. Further, according to market research, the air conditioner market in Pakistan is expected to grow by around 7% during 2017–2023. Energy-efficient HVAC systems would help in reducing the total electricity consumption as well as emissions.	Minimum Energy Performance Standard (MEPS) for Window Type and Split Air-Conditioners with Cooling Capacity under 14,000 W – issued by Ministry of Water and Power (2013) MEPS and Labeling (Energy Efficiency Standards and Labeling: ES&L) Policy/ Guidelines for Implementation of ES&L Scheme - issued by Ministry of Water and Power (2014)
District Heating and Cooling Systems	A system that can distribute heating and cooling from a centralized location through a network of pipes to communities and other establishments	As the cooling demand for the country is increasing manifold, district systems coupled with renewable energy sources for powering these systems can be looked upon as a viable solution.	

[a] Z. Ebrahim. 2021. China's coal exit will not end Pakistan's reliance on dirty fuel. *The Third Pole*. 26 October.
[b] Z. Ebrahim. 2023. Coal returns to the China–Pakistan Economic Corridor. *China Dialogue*. 13 April.
[c] Power Technology. 2023. Pakistan inaugurates third unit of Karachi Nuclear Power Plant. 3 February.
[d] R. Asghar et al. 2022. Wind energy potential in Pakistan: a feasibility study in Sindh Province. *Energies*. 15(22). p. 8333.
[e] Build Pakistan.
[f] International Water Management Institute. 2019. Solar Irrigation for Agricultural Resilience In South Asia (SoLAR): Proposal document.

Source: Compiled by authors.

CHAPTER 2

RESULTS OF THE MODEL

This section examines the results of the model and particularly the low-carbon scenario, which defines the future pathway. The main comparisons between the BAU and low-carbon scenarios are highlighted to illustrate the potential impact of actively pursuing a low-carbon strategy.

Total Final Energy Demand

Under the BAU scenario, coal, natural gas, and oil continue to be the dominant fuels in final energy demand until 2050. In the BAU scenario, total final energy demand (TFED) increases from 87 million tons of oil equivalent (Mtoe) in 2020 to 416 Mtoe by 2050, growing at a compound annual growth rate (CAGR) of 5.4%. Figure 1 shows the trends in fuel-wise TFED from 2020 to 2050.[2] Coal becomes the dominant source of supply by around 2040 and its share grows up to 2050.

Figure 1: Total Final Energy Demand by Fuel Type Under a Business-as-Usual Scenario

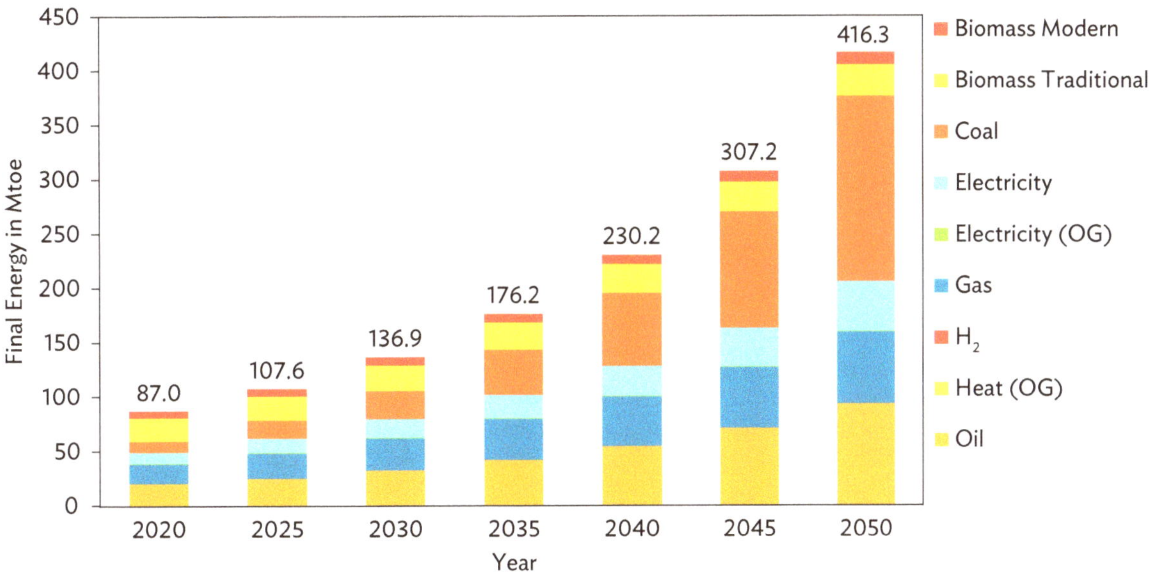

H_2 = hydrogen, Mtoe = million tons of oil equivalent, OG = off-grid.
Source: Authors.

[2] Modern biomass includes advanced, efficient biomass, and biogas-based cooking technologies.

In the low-carbon scenario, TFED would reach 127 Mtoe by 2030 and 365 Mtoe by 2050, with an estimated CAGR of 4.9% between 2030 and 2050. This means that by 2050 under low-carbon, TFED will be 12.3% below that in the BAU scenario. The key factors responsible for the reduction in TFED are the assumptions regarding fuel switching (coal-to-gas in industries) and electrification across sectors, e.g., electric pumps in agriculture, electric vehicles in the transport sector, and electric cooking. This also results in an increase in demand for electricity, which substitutes for the direct fossil-based energy requirement across various sectors. Additionally, there is a shift from coal to natural gas in the industry sector. Figure 2 shows the TFED over the period by fuel type for the low-carbon scenario.

Figure 2: Total Final Energy Demand by Fuel Type Under a Low-Carbon Scenario

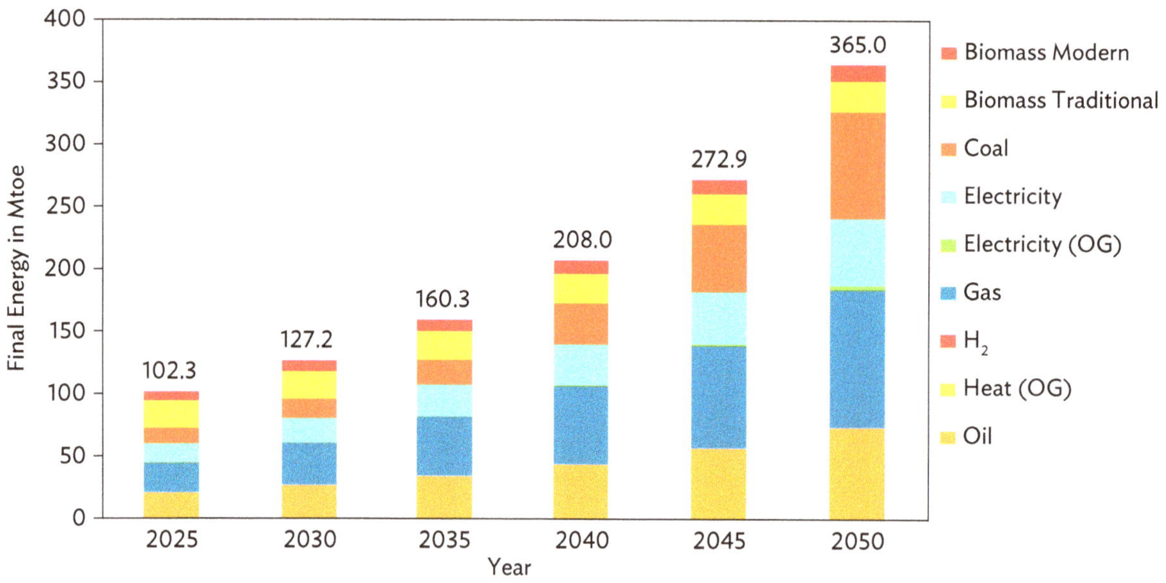

H_2 = hydrogen, Mtoe = million tons of oil equivalent, OG = off-grid.
Source: Authors.

In terms of the growth of different fuels under the low-carbon scenario, the fastest annual growth is for off-grid electricity at around 13%, followed by coal (9%) and gas (6%) (Table 5). Hence, although coal takes a smaller share in TFED than in the BAU scenario, coal use continues to grow despite no new investment in coal-based capacity installation post-2030. Coal demand post-2030 mainly results from industries, including hard-to-abate sectors associated with growth in industrial outputs, to achieve the desired economic growth and the preexisting coal-based power generation capacity.

Table 5: Compound Annual Growth Rate of Final Energy Demand by Major Fuels from 2030–2050 Low-Carbon Scenario

Coal	Oil	Gas	Biomass	Electricity Grid	Electricity Off-Grid
9%	5.1%	6.1%	1.1%	5.1%	13.4%

Source: Authors.

Figure 3 gives the mix of TFED in terms of fuel types between 2025 and 2050. The share of biomass, coal, and oil would be reduced from 67% in 2020 to 54% in 2050. On the other hand, the cumulative share of gas and electricity would increase from 33% in 2020 to 46% by 2050.

Figure 3: Fuel-Wise Share of Final Energy Demand Under a Low-Carbon Scenario

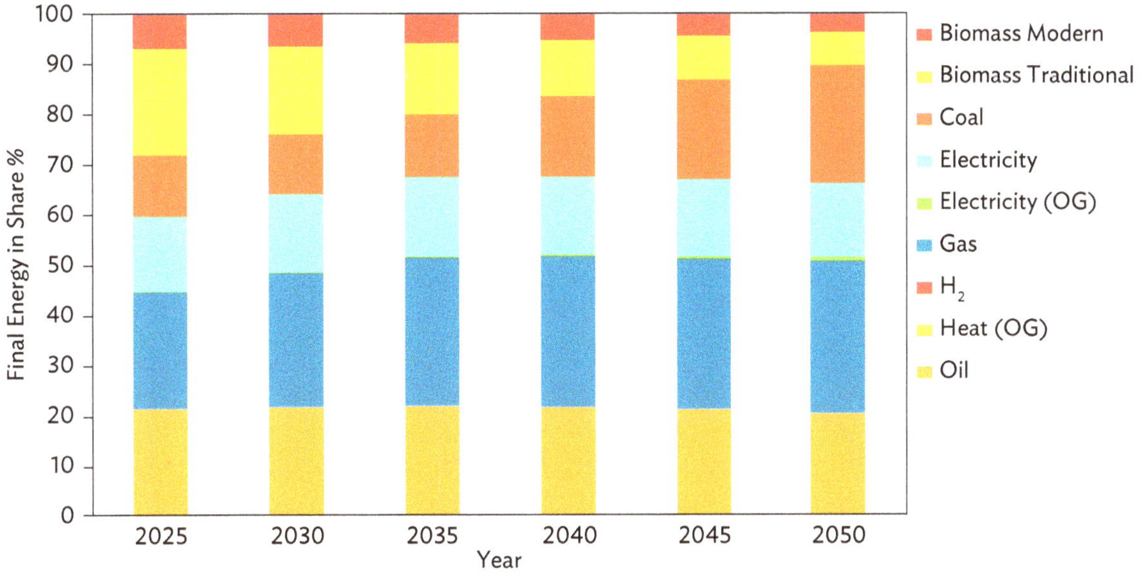

H_2 = hydrogen, OG = off-grid.
Source: Compiled by authors.

Table 6 gives the growth of energy consumption by user sectors. Under both scenarios, growth is highest in industry, which remains the largest user sector (taking 60% of total use in low-carbon and 58% in BAU in 2050). While overall there is a 12.3% reduction in the final energy demand under the low-carbon scenario compared to the BAU scenario by 2050, there are differences between user sectors. Final energy use in agriculture increases in the low-carbon scenario, as the transition toward off-grid pumping would increase the penetration of off-grid electricity use and the application of decentralized electricity in commercial and residential non-cooking applications is also assumed to increase with time. The industry sector experiences a decline in energy demand as natural gas replaces some coal and oil, although the share of coal remains significant in the fuel mix. In transport, the transition from internal combustion engine-based to electric vehicles, particularly in passenger traffic, leads to a reduction in energy demand. Transport remains the second-largest user sector with approximately 20% of demand in 2050 under the low-carbon scenario.

Table 6: Annual Growth of Energy Consumption by User Sector—Both Scenarios

Demand Sectors	% Annual Growth 2030–2050	
	Business-as-Usual Scenario	Low-Carbon Scenario
Agriculture	1.6	2.9
Commercial	6.8	6.9
Industry	7.6	7.3
Residential[a]	1.7	1.4
Transport[b]	5.0	4.4
Overall	5.7	5.4

[a] Including both cooking and appliances.

[b] Including both passenger and freight transport.

Source: Compiled by authors.

Figure 4 shows the quantity of TFED by different sectors, under the low-carbon scenario and Figure 5 gives the percentage breakdown.

Figure 4: Total Final Energy Demand by Sector Under a Low-Carbon Scenario

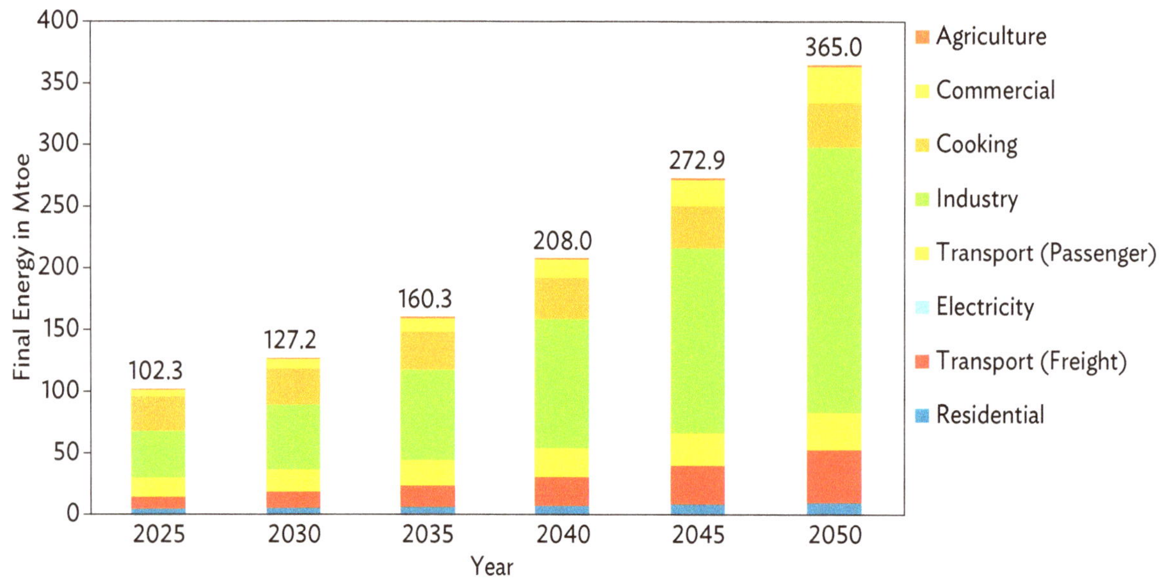

Mtoe = million tons of oil equivalent.

Source: Authors.

Figure 5: Sector-Wise Share of Final Energy Demand Under a Low-Carbon Scenario

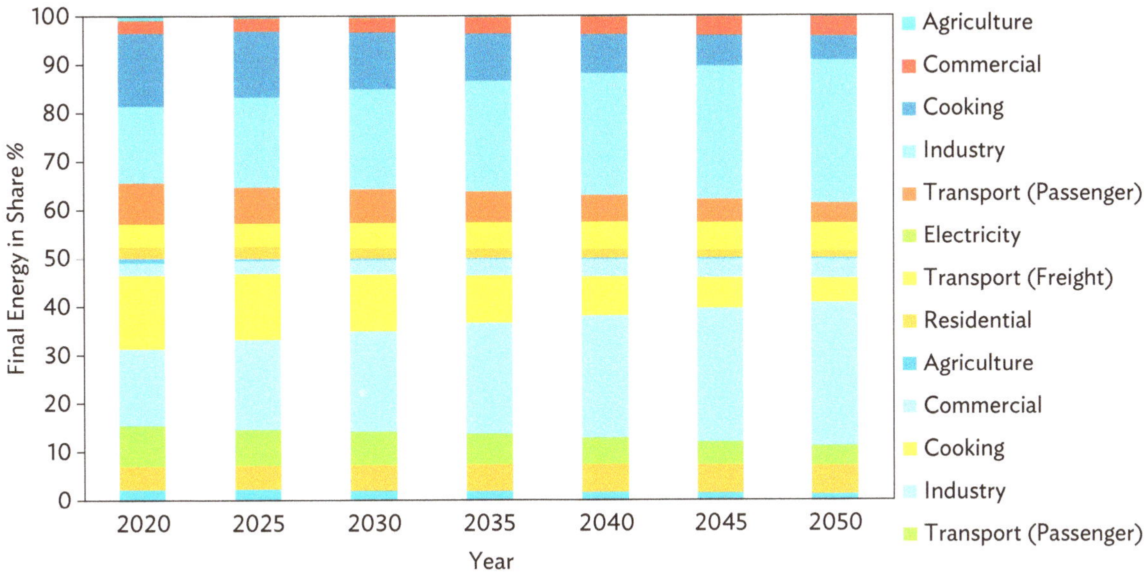

Source: Authors.

Total Primary Energy Supply

Historically, total primary energy supply (TPES) has been dominated by oil, natural gas, and biomass. Between 1990 and 2019, the share of oil decreased from 25% in 1990 to 20%, while the share of natural gas increased from 23% in 1990 to 31%. Biomass has been important due to the heavy dependence of the rural population on traditional biomass for cooking and heating. Its share of primary energy supply (PES) declined from 44% in 1990 to 33% in 2019 due to the increasing role of imports of oil and natural gas. A significant shift has been the increasing role played by coal, whose share rose from 5% in 2015 to 12% in 2019 due to the energy policy of diversifying to imports of coal and increasing domestic coal production. Currently, the role of renewables remains small and comprised only 0.3% in 2019.

Under the BAU scenario, the TPES in Pakistan would reach 159.7 Mtoe by 2030 and 491.9 Mtoe by 2050, growing at a CAGR of 5.8%. Figure 6 shows the trends in the domestic PES. Coal would account for the maximum share at of around 15% in 2030, rising to 74% by 2050. The sudden growth in domestic coal production post-2040 is primarily due to the depleting domestic natural gas reserves, which resulted in additional demand for coal under the cost-optimization results of the model. The total domestic PES would reach 93.9 Mtoe by 2030 and 229.5 Mtoe by 2050, growing at a CAGR of 4.6%.

Figure 6: Fuel-Wise Domestic Primary Energy Supply for Pakistan Under a Business-as-Usual Scenario

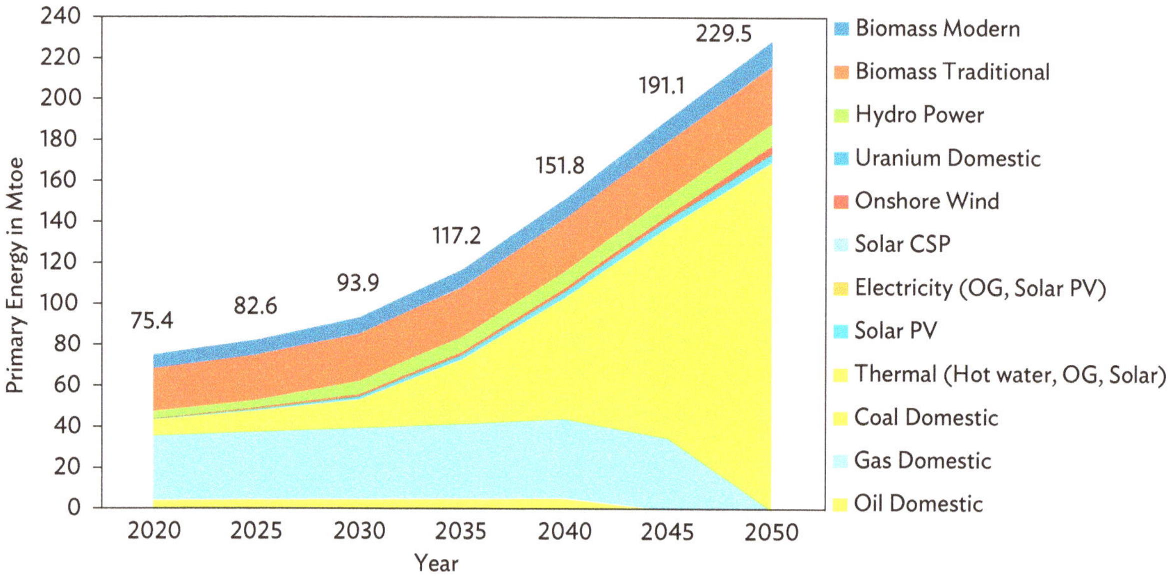

CSP = concentrated solar power, Mtoe = million tons of oil equivalent, OG = off-grid, PV = photovoltaic.

Source: Authors.

Figure 7 shows the trends in imported primary energy, which would reach 65.84 Mtoe by 2030 and 262.5 Mtoe by 2050, with a CAGR of 7.2%. The import dependency (imports in TPES) would increase from 41% in 2030 to 53% in 2050.

Figure 7: Primary Energy Import by Source in Pakistan Under a Business-as-Usual Scenario

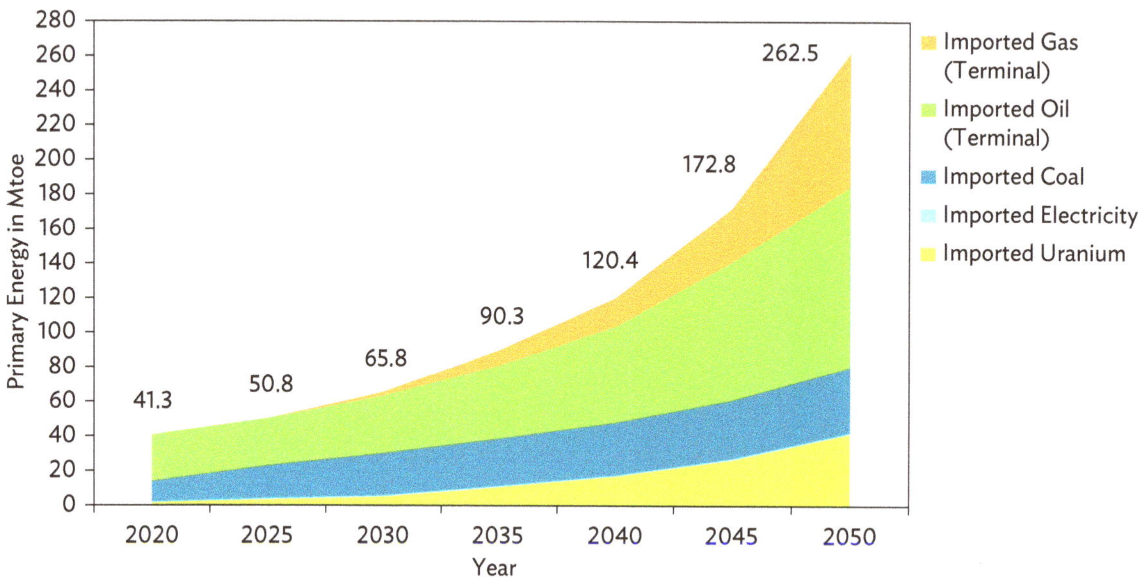

Mtoe = million tons of oil equivalent.

Source: Authors.

Table 7 shows the CAGR of energy supply by domestic fuel commodities between 2030 and 2050 under a BAU scenario. Coal rises with the maximum CAGR of 13.3% between 2030 and 2050 among all other fuels while contribution of domestic oil and gas in domestic primary energy diminishes to zero by 2050.

Table 7: Compound Annual Growth Rate of Domestic Primary Energy Supply by Various Fuel Commodities from 2030–2050 Under a Business-as-Usual Scenario

Biomass Traditional	Biomass Modern	Coal	Oil	Gas	Solar	Onshore Wind
2.2%	1%	13.3%	(100%)	(100%)	5.5%	7.7%

( ) = negative.
Source: Compiled by authors.

Table 8 provides the growth rate of energy imports for various fuels between 2030 and 2050 under the BAU scenario. The highest increase is for gas and uranium, in line with the limited domestic availability of natural gas. Further, with increase in nuclear-based electricity generation, uranium imports would grow at a CAGR of 10.4% between 2030 and 2050.

Table 8: Compound Annual Growth Rate of Energy Import by Fuel Type from 2030–2050 Under a Business-as-Usual Scenario

Coal	Gas	Oil	Uranium
%	19.6%	5.9%	10.4%

Source: Compiled by authors.

Under the low-carbon scenario, TPES is estimated to reach 155.6 Mtoe in 2030 and 429.3 Mtoe in 2050 at a CAGR of 5.2%. The domestic contribution toward TPES is shown in Figure 8. Domestic PES would reach 94.5 Mtoe by 2030 and 165.9 Mtoe by 2050, with a CAGR of 2.9%.

Figure 8: Primary Energy Supply by Domestic Fuel Sources Under a Low-Carbon Scenario

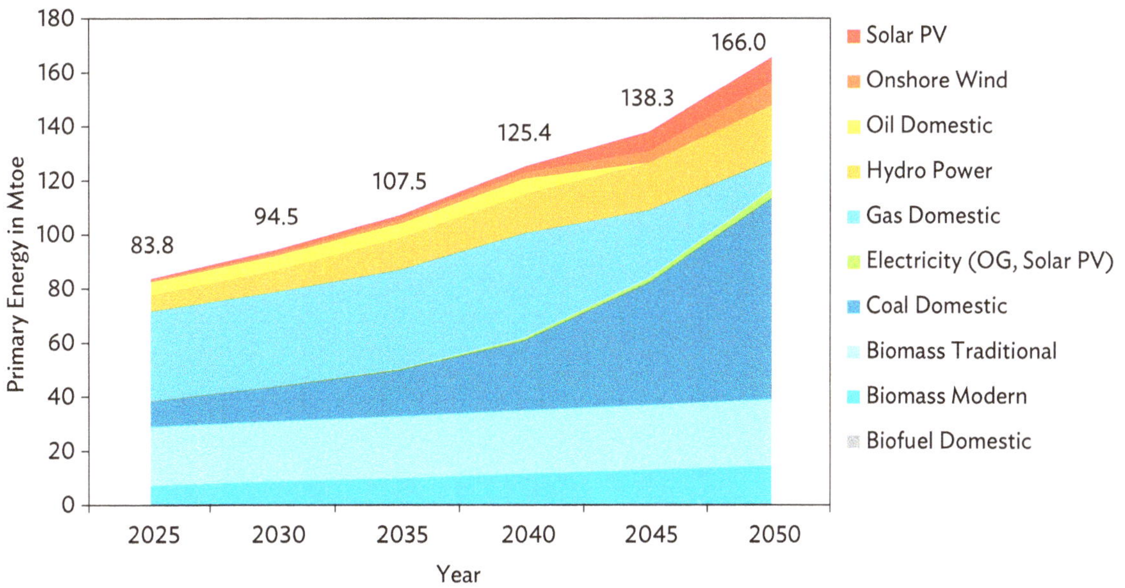

Mtoe = million tons of oil equivalent, OG = off-grid, PV = photovoltaic.
Source: Authors.

Table 9 shows the CAGR of energy supply by domestic fuel commodities between 2030 and 2050, with strong annual growth for wind, solar, and coal.

Table 9: Compound Annual Growth Rate of Domestic Primary Energy Supply by Various Fuel Commodities from 2030–2050 Under a Low-Carbon Scenario

Biomass Traditional	Biomass Modern	Coal	Oil	Gas	Solar	Onshore Wind
0.5%	2.5%	9.2%	(100%)	(5.9%)	9.7%	12.2%

( ) = negative.
Source: Compiled by authors.

Figure 9 shows the PES by imports. Energy imports are estimated to reach 61.1 Mtoe by 2030 and increase to 263.3 Mtoe by 2050 at a CAGR of 5.2%. The share of imports in the TPES would be 39% in 2030 and would reach 61% by 2050, which is higher than in the BAU scenario. It is worthwhile to note that coal continues to play a role even under the low-carbon scenario despite no new investment in coal-based power plants post-2030 because of demand from preexisting power plants and industries contributing to the intended socioeconomic development in the country.

Figure 9: Primary Energy Supply by Imports Under a Low-Carbon Scenario

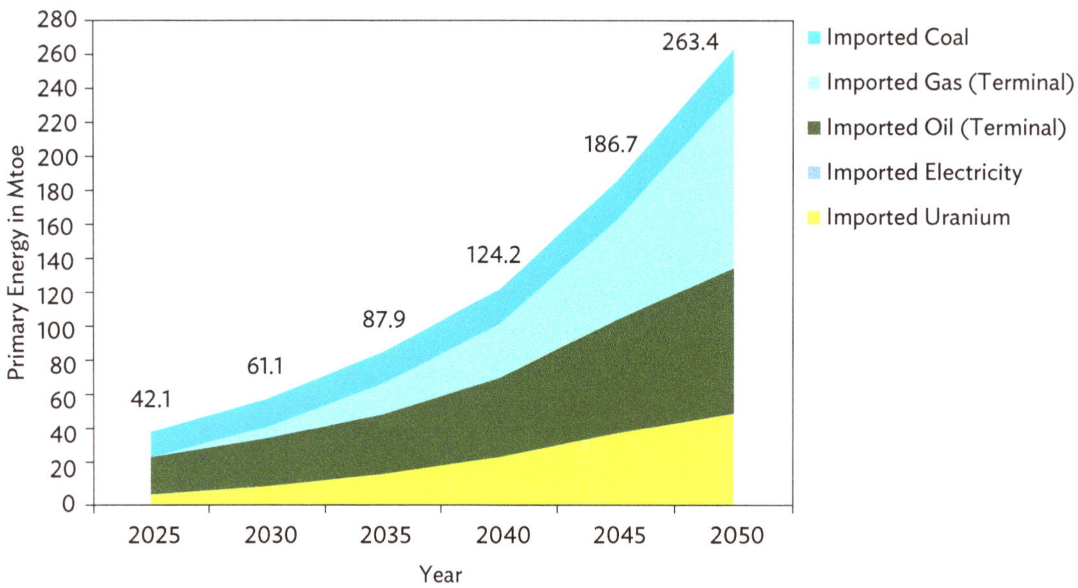

Mtoe = million tons of oil equivalent.
Source: Authors.

Table 10 gives the growth rate of energy imports for various fuels between 2030 and 2050. The highest increase is for gas and uranium, in line with the diminishing supply of domestic natural gas. Further, due to the electricity supply by nuclear power plants, uranium would be imported at an estimated CAGR of 8.3% between 2030 and 2050, lower than the BAU scenario. However, it should be noted that the growth in nuclear-based power generation is higher and begins early in the low-carbon scenario, therefore the lower growth rate in the last 2 decades compared to the BAU scenario. The growth in imported uranium between 2020 and 2050 under low-carbon scenario is 10.8%, compared to 10% under the BAU scenario.

Table 10: Compound Annual Growth Rate of Energy Import by Fuel Type from 2030–2050 Under a Low-Carbon Scenario

Coal	Gas	Oil	Uranium
2.2%	14.6%	5.7%	8.3%

Source: Compiled by authors.

Figure 10 and Figure 11 show the fuel-wise composition of the TPES between 2025 and 2050 under the low-carbon scenario. The country would continue to depend on natural gas (including domestic and import) as a major source of energy with an almost constant share of about 26% in 2030 and 2050. Renewables like hydro, wind, and solar power in total would supply 7% in 2030 and 10% in 2050.

Figure 10: Total Primary Energy Supply by Fuel Under a Low-Carbon Scenario

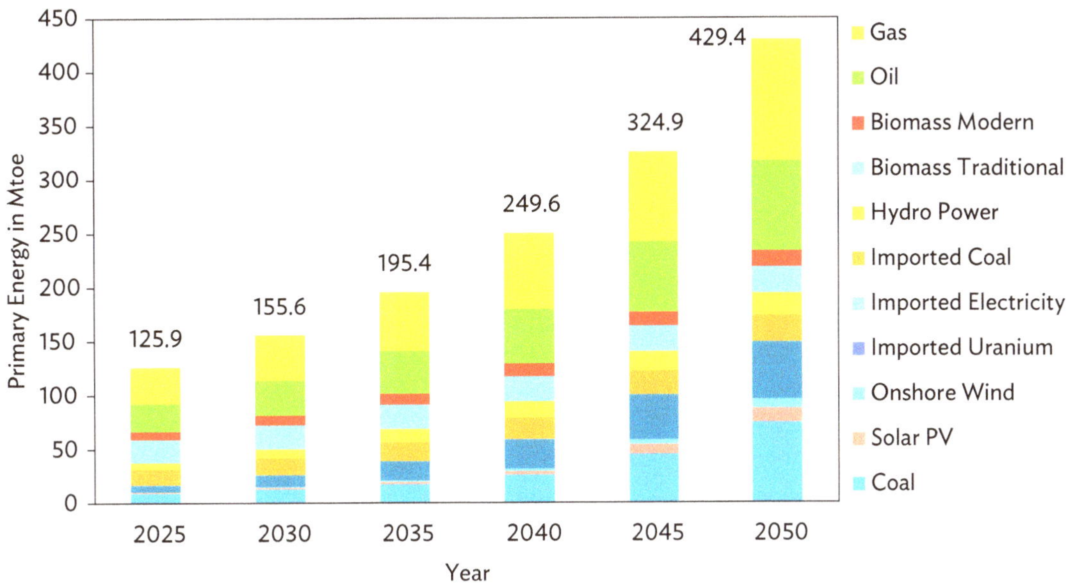

Mtoe = million tons of oil equivalent, PV = photovoltaic.

Source: Authors.

The share of oil would drop slightly from 21% in 2030 to 20% in 2050. Coal's share in the TPES would remain large at 23% in 2050 (17% from domestic production and 6% from imports). This contrasts with a 43% share in the BAU scenario (35% from domestic production and 8% from imports). Traditional biomass would still be used due to demand in the residential sector, though its share would reduce significantly from 14% in 2030 to 6% in 2050.

Figure 11: Share of Total Primary Energy Supply by Fuel Under a Low-Carbon Scenario

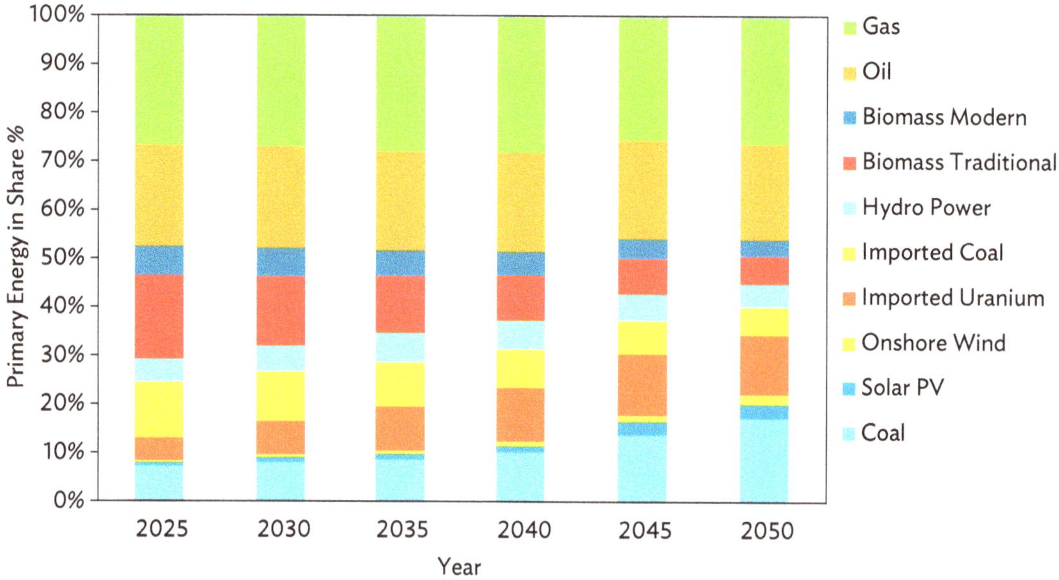

PV = photovoltaic.
Source: Authors.

Although the low-carbon scenario results in lower energy requirements than the BAU scenario, import dependence (imports in total primary energy supply) in 2050 would be 61% as compared to 53% in the BAU scenario. Although, in absolute terms, the low-carbon scenario would require an almost similar quantum of energy imports, the imported fuel composition would be different. In the BAU scenario, by 2050, oil imports would account for the highest share, about 40% of total energy imports followed by gas (29%) and coal (14%). Under the low-carbon scenario, the highest growth in energy imports would be in natural gas as the country transitions toward natural gas, nuclear, and renewables as the key energy sources. By 2050, natural gas would account for the largest share of imports, about 40% of total energy imports, followed by oil (32%) and uranium (20%). Energy security planning in the country would therefore need to account for this increased import dependence in different types of primary energy, with its associated geopolitical risks to energy security.

Structure of Electricity Generation

Electricity generation in Pakistan had been dominated by oil and natural gas-fired power plants, whose combined contribution was around 56% of total generation in 2019. The share of electricity generation from coal-fired plants had been low historically. This increased after 2015, reaching 12% in 2019. The share of hydro-based electricity fell from 45% in 1990 to 21% in 2019. In 2019, the share of renewable technologies like wind and nuclear power was relatively low at 8% for nuclear and 3% for wind in 2019. The absolute amount of electricity generation in 1990–2019 by fuel is shown in Figure 12.

Figure 12: Electricity Generation Mix for Pakistan, 1990–2019

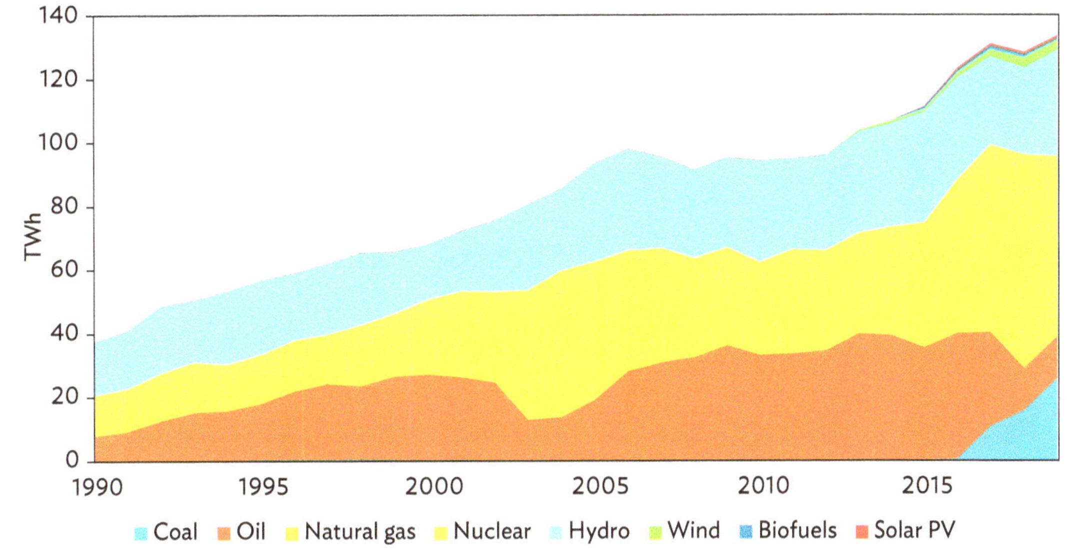

PV = photovoltaic, TWh = terawatt-hour.
Source: International Energy Agency, Pakistan. 2021.

In the BAU scenario, the total electricity generation would grow from 243.4 terawatt-hours (TWh) in 2030 to 630 TWh in 2050, an annual CAGR of 4.9%. The source-wise trend in electricity generation is in Figure 13.

Figure 13: Electricity Generation by Source in Terawatt-Hours Under a Business-as-Usual Scenario

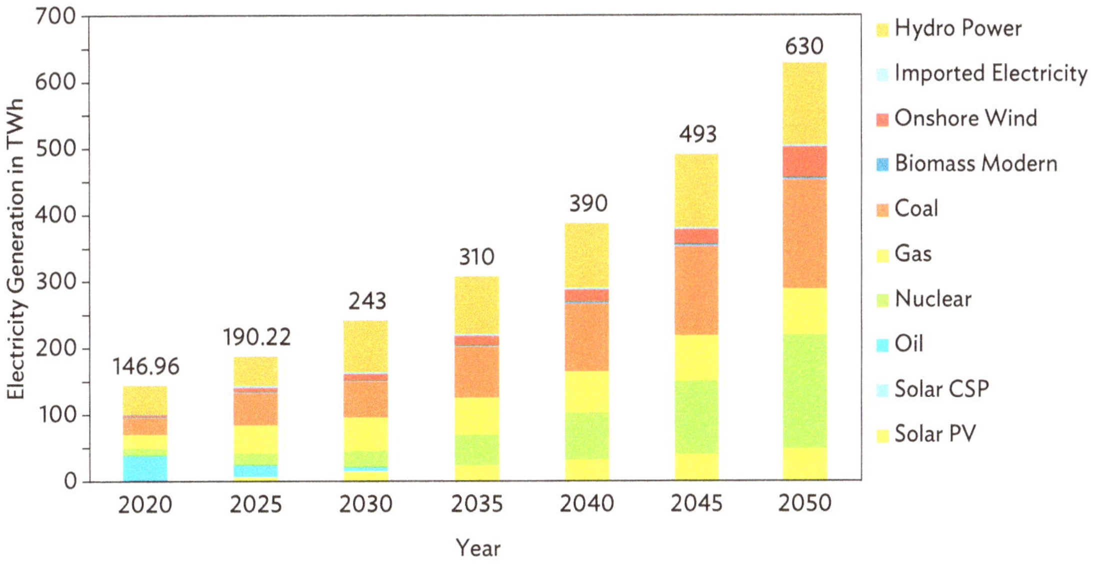

CSP = concentrated solar power, PV = photovoltaic, TWh = terawatt-hour.
Source: Authors.

The total installed capacity would grow from 61.2 gigawatts (GW) in 2030 to 155 GW in 2050, a CAGR of 4.8%. The trend in installed capacity is presented in Figure 14. The BAU scenario allows for a growth of renewable sources, particularly onshore wind, but also a relatively rapid growth of power from coal sources, using coal supercritical technology-based power plants.

Figure 14: Cumulative Installed Electricity Capacity by Source in Gigawatts Under a Business-as-Usual Scenario

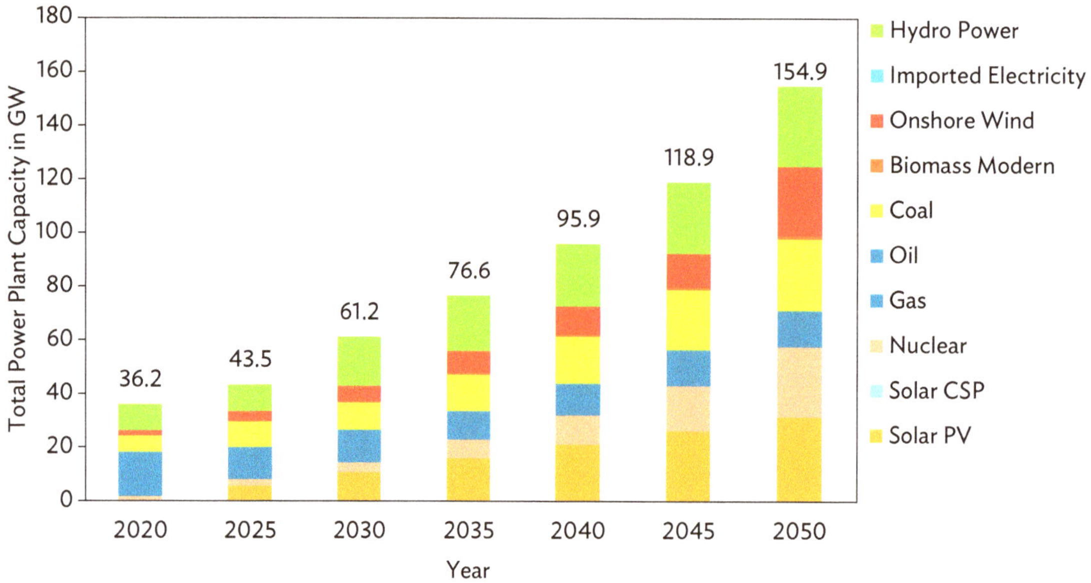

CSP = concentrated solar power, GW = gigawatt, PV = photovoltaic.
Source: Authors.

Under the low-carbon scenario, total electricity generation would reach 275.7 TWh by 2030 and 729.2 TWh by 2050, a CAGR of 5% (Figure 15). Hydropower and nuclear would remain the two largest sources of electricity generation, but the share of coal would drop and that of renewables increase strongly.

Figure 15: Electricity Generation by Source in Terawatt-Hours Under a Low-Carbon Scenario

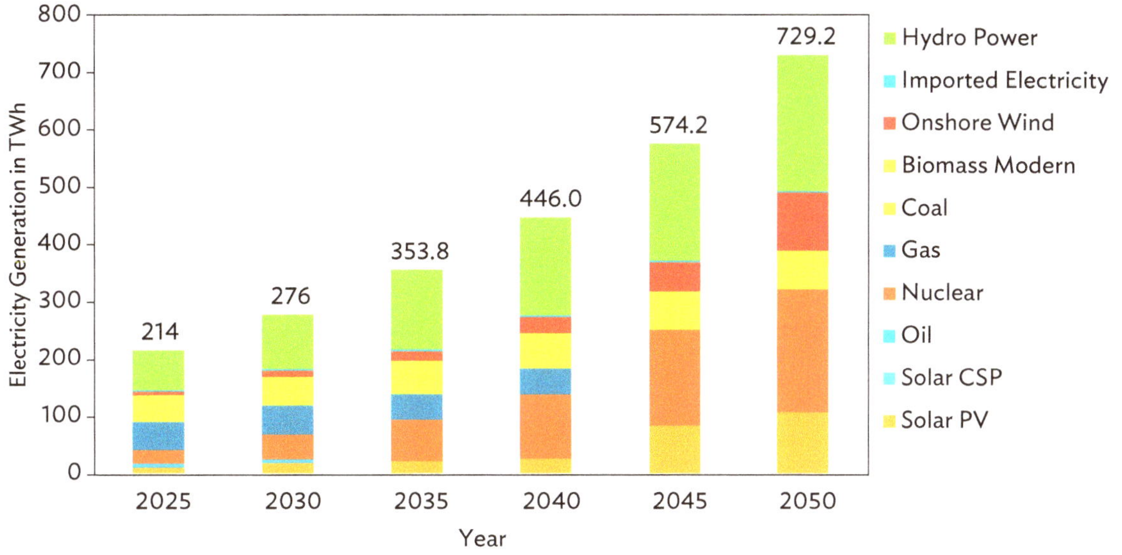

CSP = concentrated solar power, PV = photovoltaic, TWh = terawatt-hour.
Source: Authors.

A detailed breakdown in terms of share between 2025 and 2050 is in Figure 16. Non-fossil sources of energy, including nuclear and hydro, would together account for a 60% share in electricity generation by 2030 and 91% in 2050.

Figure 16: Source-Wise Share in Electricity Generation Under a Low-Carbon Scenario

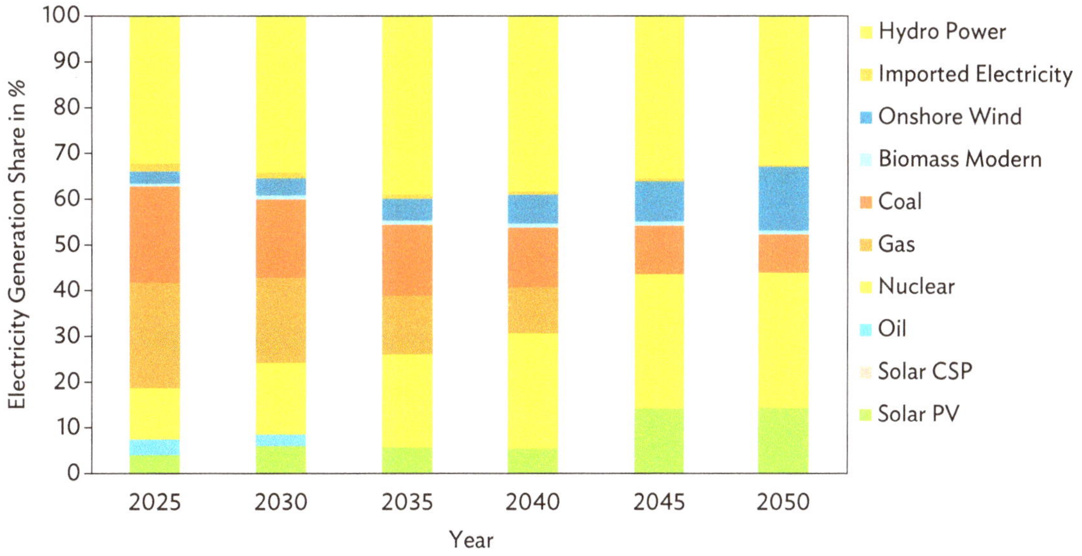

CSP = concentrated solar power, PV = photovoltaic.
Source: Authors.

Reflecting current policy objectives, there would be a significant increase in electricity generation from clean energy sources such as solar PV and hydropower. In 2050, electricity generation by hydropower would account for 32% of the total as compared with 19% in the BAU scenario. Generation by solar PV and wind onshore plants would be 29% of the total, as compared with 15% in BAU. Nuclear-based electricity would be 29% of the total as compared with 27% in BAU. On the other hand, coal-based generation under the low-carbon scenario is 8% of the total, as compared with 26% in BAU.

Figure 17 shows the trend in electricity installed capacity under the low-carbon scenario. The total installed capacity is 67.1 GW in 2030 and 234.1 GW in 2050, a CAGR of 6%.

Figure 17: Cumulative Installed Electricity Capacity by Source in Gigawatts Under a Low-Carbon Scenario

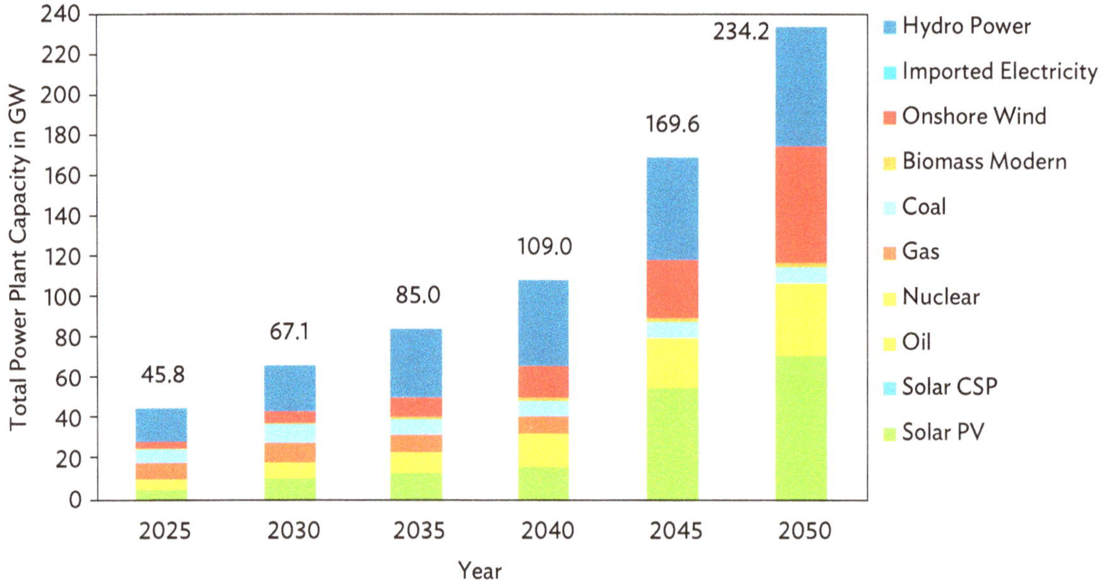

CSP = concentrated solar power, GW = gigawatt, PV = photovoltaic.

Source: Authors.

The source-wise breakdown in terms of share of installed capacity is shown in Figure 18. Renewables, including solar PV, onshore wind, and hydropower in total account for 40.2 GW in 2030 (60% of the total) in 2030, rising to 190.3 GW in 2050 (81% of the total), a CAGR of 8% between 2030 and 2050.

Figure 18: Source-Wise Share in Installed Electricity Capacity in Gigawatts Under a Low-Carbon Scenario

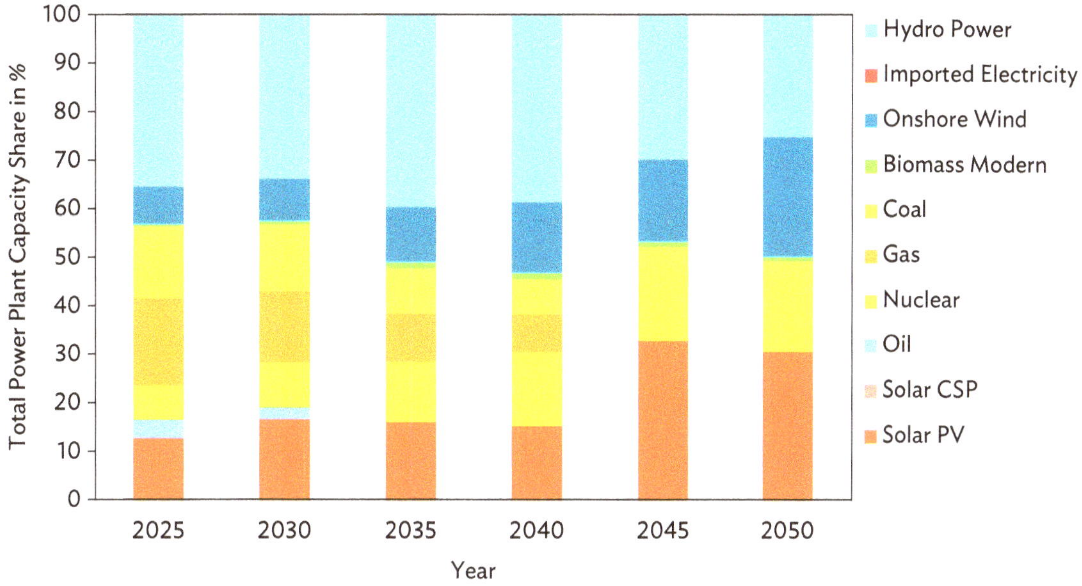

CSP = concentrated solar power, PV = photovoltaic.
Source: Authors.

Analysis of Greenhouse Gas Emissions

While low on a per capita basis, emissions from energy-related sources have more than doubled since 2000, from 94.4 million tons of carbon dioxide (tCO_2) to 183.4 million tCO_2 in 2017, a CAGR of 4%. At the sector level, power, followed by industry and transport, are the main sources of emissions. Carbon dioxide (CO_2) emissions from the power sector increased from 33 million tCO_2 in 2000 to 55 million tCO_2 at a CAGR of 3%.[3] Emissions reduction should be an important benefit from the pursuit of a low-carbon strategy and the potential reduction can be seen from a comparison of the BAU and low-carbon emissions profiles.

Total GHG emissions in terms of carbon dioxide equivalent (CO_2e) are estimated for CO_2, methane, and nitrous oxide. For the BAU scenario, Figure 19 shows the total GHG emissions from the energy supply sector. The emissions from the supply side grow at a CAGR of 5% from 2030 to 2050. The total GHG emissions from the supply side would increase from 81.8 million tons of carbon dioxide equivalent per year ($MtCO_2e/yr$) in 2030 to 212.3 Mt CO_2e/yr in 2050, predominantly from coal-based energy (Figure 20).

[3] IEA. 2019. CO_2 Emissions from Fuel Combustion 2019 (accessed 20 January 2023).

Figure 19: Total Supply-Side Greenhouse Gas Emissions by Energy Sector Under a Business-as-Usual Scenario

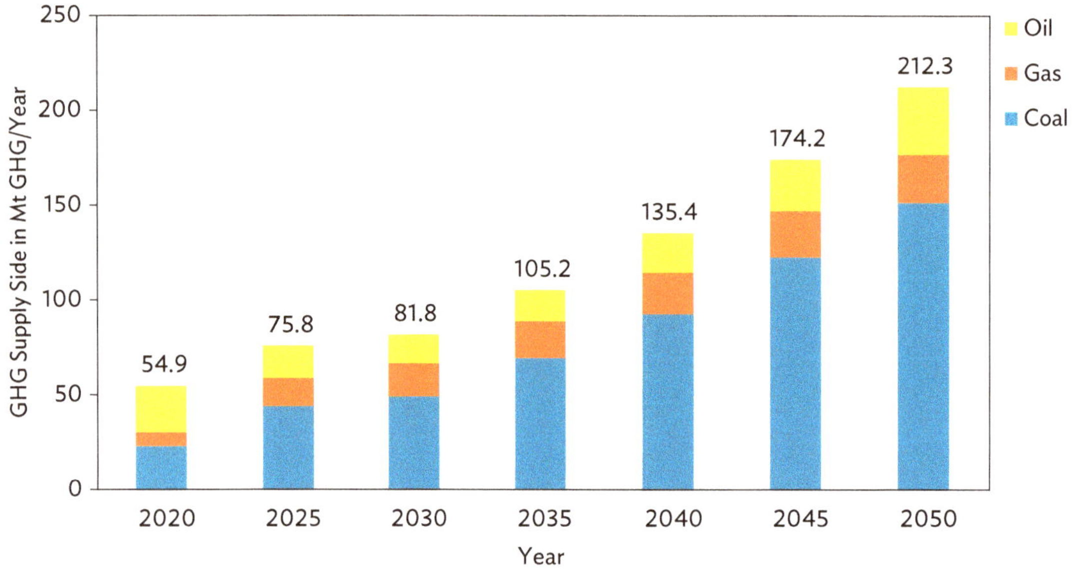

GHG = greenhouse gas, Mt = million tons.
Source: Authors.

Figure 20: Fuel-Wise Share of Supply-Side Greenhouse Gas Emissions Under a Business-as-Usual Scenario

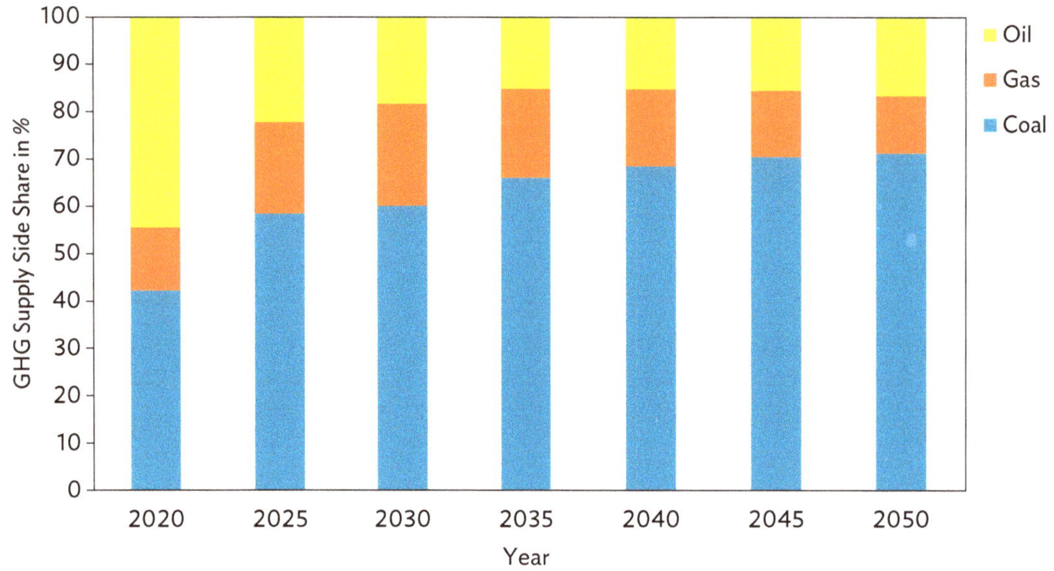

GHG = greenhouse gas, Mt = million tons.
Source: Authors.

Figure 21: Total Supply-Side Greenhouse Gas Emissions from Electricity Generation Under a Business-as-Usual Scenario

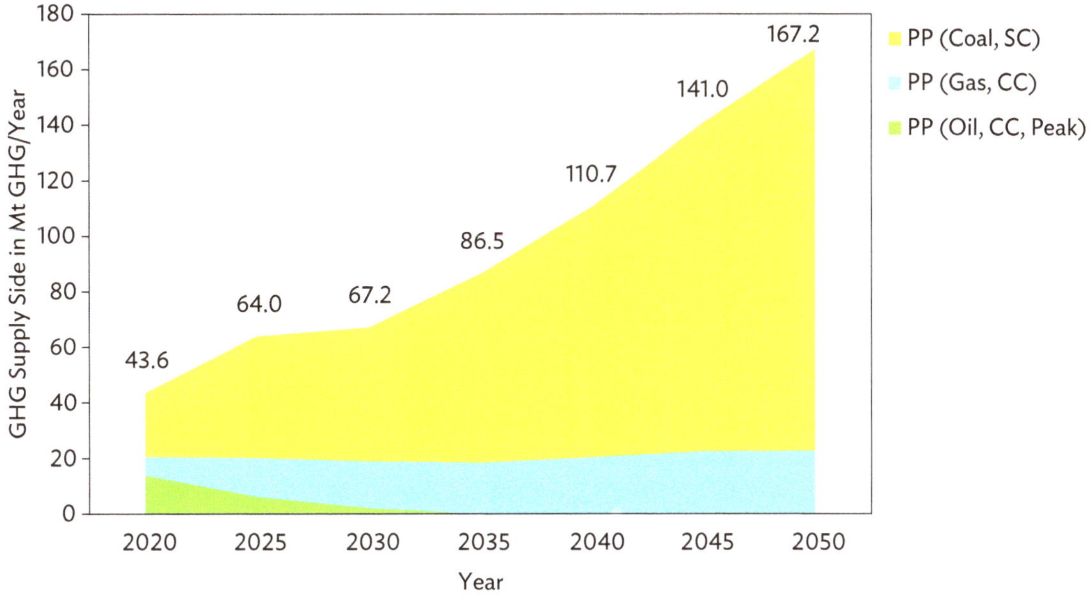

CC = combined cycle, GHG = greenhouse gas, Mt = million tons, PP = power plant, PV = photovoltaic, SC = super critical.
Source: Authors.

The total GHG emissions from electricity generation would be 67.2 $MtCO_2e$/yr in 2030, and 167.2 $MtCO_2e$/yr in 2050. By 2050 coal power plants would emit more than 86%, with the remainder from gas power plants (Figure 21).

On the demand-side, users in total would emit 333.8 $MtCO_2e$/yr in 2030, and 1,164.1 $MtCO_2e$/yr in 2050. As can be seen from Figure 22 and Figure 23, the industry sector would be the largest emitter accounting for 48% of emissions within the demand sectors in 2030, and 64% in 2050. The share of transport, the other major emitting sector, decreases from 46% in 2030 to 33% in 2050. Other sectors contribute only modest percentage shares.

Figure 22: Total Demand-Side Greenhouse Gas Emissions Under a Business-as-Usual Scenario

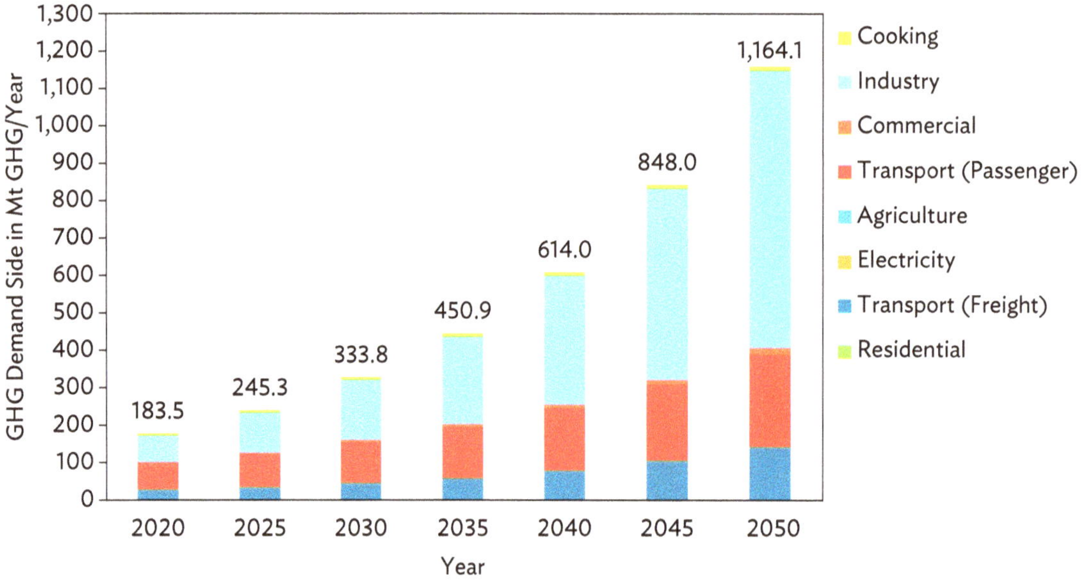

GHG = greenhouse gas, Mt = million tons.
Source: Authors.

Figure 23: Sector-Wise Share of Demand-Side Greenhouse Gas Emissions Under a Business-as-Usual Scenario

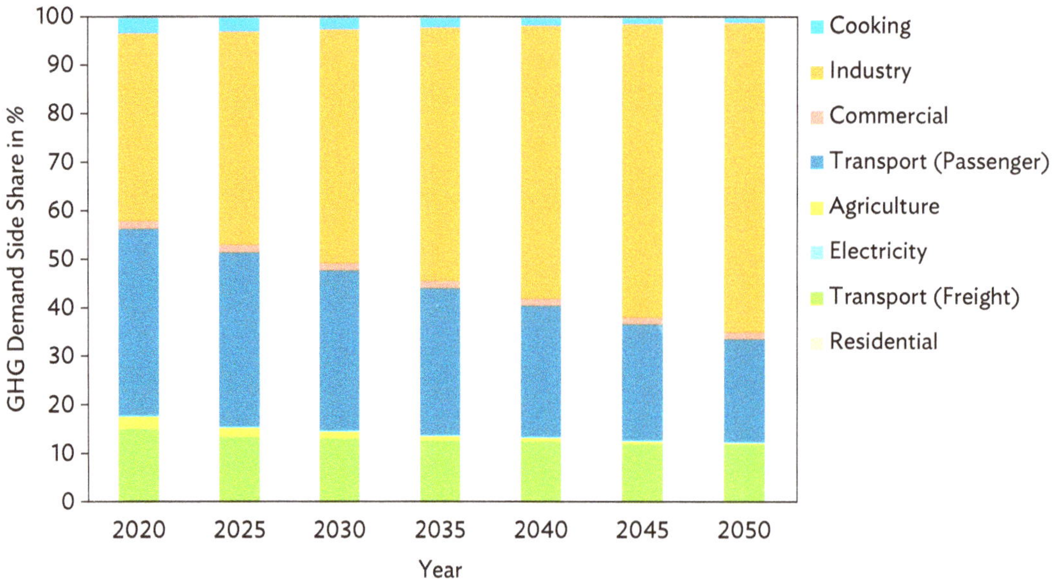

GHG = greenhouse gas.
Source: Authors.

Under the low-carbon scenario, total GHG emissions would reach 326.4 $MtCO_2e$ in 2030 and 885.1 $MtCO_2e$ in 2050 growing at a CAGR of 5% over that period. Compared to the BAU, the total GHG emissions (supply and demand sides combined) would be lower by almost 23% in 2030 and 38% by 2050.

Figure 24 gives the trend in supply-side GHG emissions from the energy sector under the low-carbon scenario. The total supply-side GHG emissions reach 73.7 $MtCO_2e$ and 89.4 $MtCO_2e$ by 2030 and 2050 respectively. Compared with the BAU, the supply-side GHG emissions under the low-carbon scenario are 10% lower in 2030 and 58% lower in 2050.

Figure 24: Total Supply-Side Greenhouse Gas Emissions from the Energy Sector Under a Low-Carbon Scenario

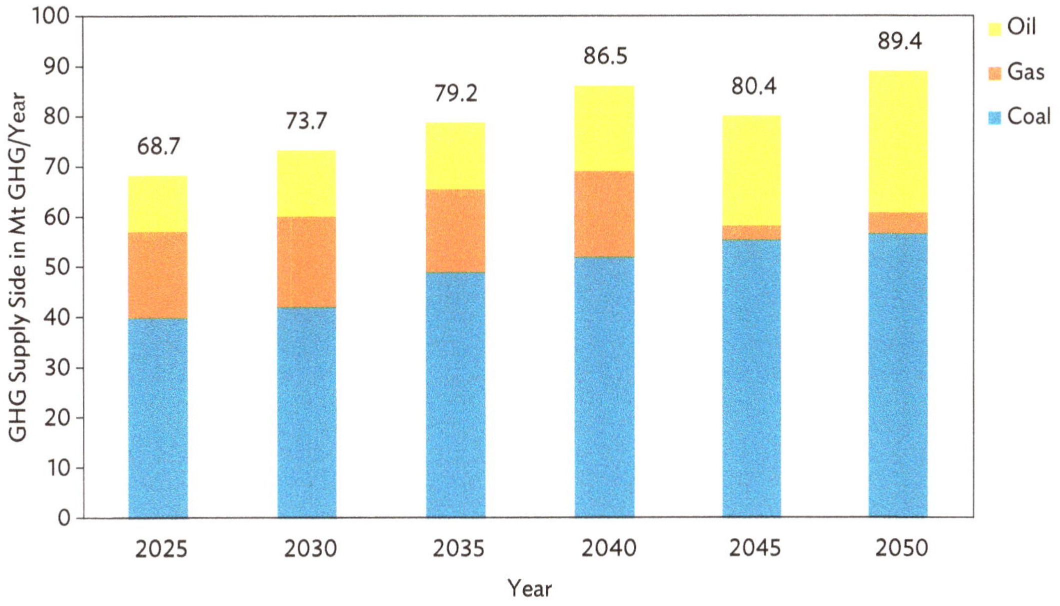

GHG = greenhouse gas, Mt = million tons.
Source: Authors.

The electricity sector would be responsible for 61.2 $MtCO_2e$ of GHG emissions in 2030 and a further 53.4 $MtCO_2e$ in 2050. Of these, coal-based emissions account for 68% of emissions in 2030, while coal accounts for the entire emissions from the electricity supply sector by 2050 (Figure 25).

Figure 25: Total Supply-Side Greenhouse Gas Emissions from Electricity Generation Under a Low-Carbon Scenario

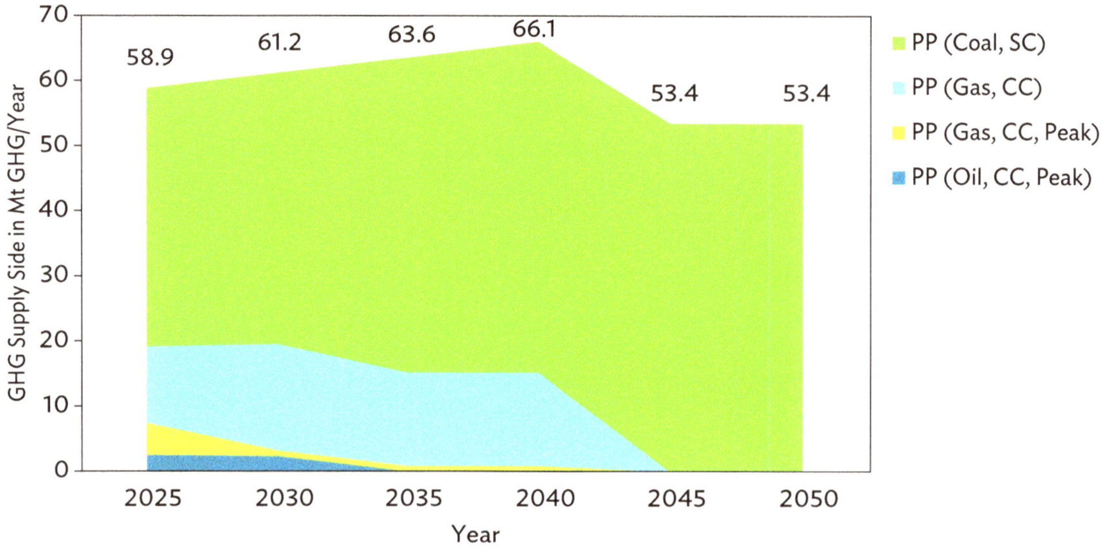

CC = combined cycle, GHG = greenhouse gas, Mt = million tons, PP = power plant, PV = photovoltaic, SC = super critical.
Source: Authors.

Under the low-carbon scenario, GHG emissions from the demand-side reach approximately 253.4 $MtCO_2e$ in 2030 and 792.6 $MtCO_2e$ by 2050. Compared with BAU, the demand-side GHG emissions are 24% lower in 2030 and 32% lower in 2050 (Figure 26).

Figure 26: Total Demand-Side Greenhouse Gas Emissions Under a Low-Carbon Scenario

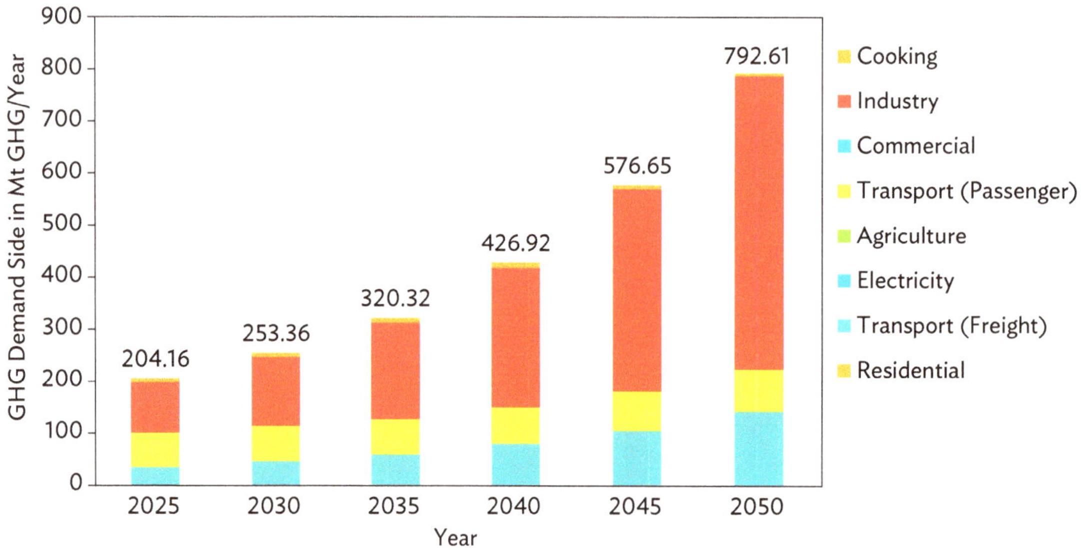

GHG = greenhouse gas, Mt = million tons.
Source: Authors.

Figure 27 gives the sector-wise distribution of GHG emissions from the demand sectors. Industry remains the major emitting sector with 133 $MtCO_2e$ in 2030 (53% of the total demand-side emissions) and 564.5 $MtCO_2e$ in 2050 (71% of the total). The transport sector, which includes both freight and passenger transport, has a share of 44% in 2030, and 28% in 2050.

Figure 27: Sector-Wise Share of Demand-Side Greenhouse Gas Emissions Under a Low-Carbon Scenario

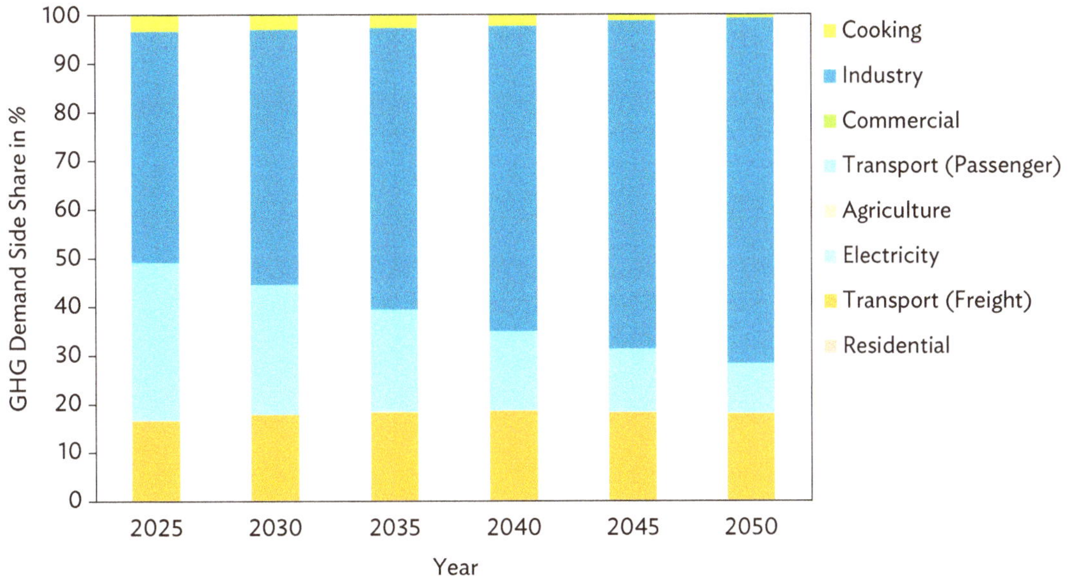

GHG = greenhouse gas.
Source: Authors.

In summary, the pathway mapped out in the low-carbon scenario, if followed, offers a way to meet the government's decarbonization objective. Under the low-carbon scenario, the power sector is transitioning toward a diverse mix of energy sources including the expansion of nuclear, solar, and hydro-based power generation. The continuation of current trends under the BAU scenario would see 37% of the total power supply coming from fossil-based power generation in 2050, with 26% from coal and 11% from natural gas. In the low-carbon scenario by 2050, around 9% of electricity would be generated from fossil fuels, largely from coal.

The low-carbon scenario shows how it is possible for the economy to grow while minimizing the effect of this growth on total GHG emissions. Comparing the low-carbon strategy with the continuation of current trends in the BAU scenario shows that emissions will be approximately one-third (36%) lower in 2050; this is a mitigation of 494 $MtCO_2e$. To achieve this requires adjustment on both the demand and supply sides. By 2050 the low-carbon scenario has a lower TFED, relative to the BAU, of 50 Mtoe. This is a 12% reduction relative to the BAU. Most of this reduction will come from transport, with the shift to vehicle electrification and industry, as coal and oil are replaced by natural gas.

In addition, energy supply will shift more rapidly to green sources under the low-carbon scenario. By 2050, the share of electricity generation from renewable energy would increase to more than 61%, with 32% from hydro, 15% from solar, and 14% from wind. However, natural gas and coal continue to play a role. Natural gas imports would have a share of 26% in TPES by 2050 owing to demand from industry and residential sectors. Also, coal will remain significant, with a share in the TPES reaching 23% in 2050 (17% domestic and 6 % imports). Additions to capacity of coal-based plants are taken to cease in 2030, so the continued growth of coal usage comes from the pre-2030 investments. Further, as Pakistan grows—aspiring to become an upper middle-income country by 2047—the demand for coal including from hard-to-abate sectors would continue. It is further assumed that any new addition in coal-based capacity by 2025 would utilize the efficient supercritical coal-based technology.

Although the low-carbon scenario results in lower energy requirements than the BAU scenario, import dependency would increase (61% as compared to 53% in the BAU scenario in 2050). Under the low-carbon scenario, the highest growth in energy imports will be in the form of natural gas and uranium. This may be an issue from an energy security perspective.

The low-carbon scenario provides the basis of a road map for the evolution of the energy sector in Pakistan on a sustainable basis. The subsequent sections discuss technology options under the strategy and the investment requirements for its implementation.

CHAPTER 3

ALTERNATIVE ENERGY SOURCES AND EFFICIENCY IMPROVEMENTS

The model focuses on seven energy supply sources and five end-user sectors (Figure 28). This section discusses some of the changes in these required for the implementation of the low-carbon scenario. It also reports the estimated emissions mitigation each can create in comparison with a fossil-fuel-based alternative.

Figure 28: Supply and Demand Sectors

Energy Supply

- Nuclear power generation
- Hydropower generation
- Gas power plant
- Wind power generation
- Solar power generation
- Biomass power generation
- Coal power plant

Energy Demand

- Buildings
- Industry
- Transport
- Cooking
- Agriculture

Source: Compiled by authors.

Nuclear

Nuclear is a zero-emissions clean energy source for electricity generation and under the low-carbon scenario it is projected to grow from 3.1 GW in 2025 to 16.5 GW in 2040 and to 36 GW in 2050 (Figure 29), when it would supply 29% of energy demand (Figure 30).

Figure 29: Nuclear Power Generation Capacity in a Low-Carbon Scenario
(GW)

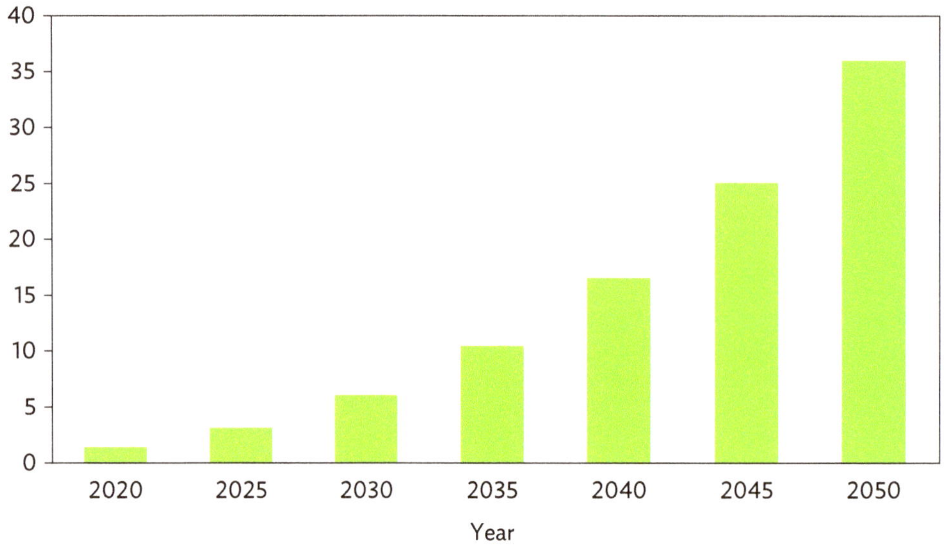

GW = gigawatt.
Source: Authors.

Figure 30: Share of Electricity Generated from Nuclear Under a Low-Carbon Scenario
(%)

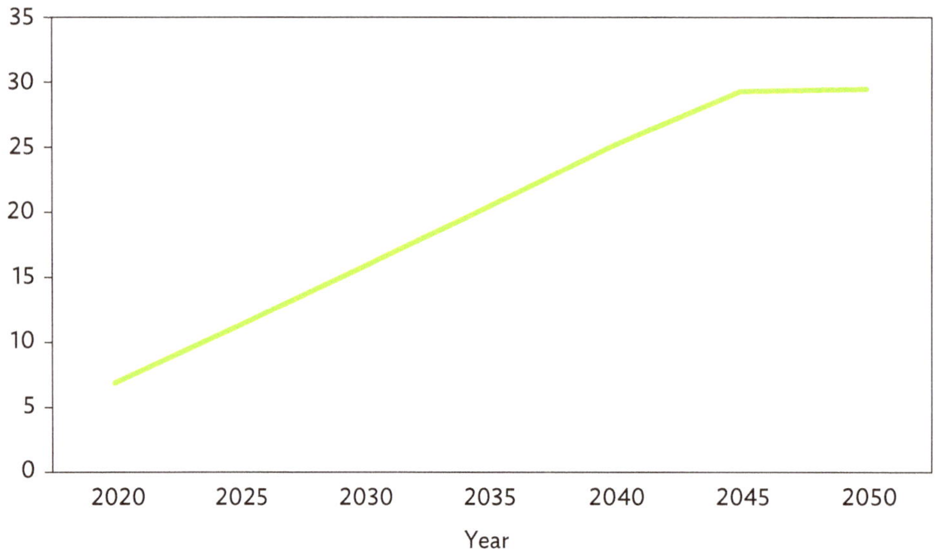

Source: Authors.

As a relatively mature technology, most nuclear harnessing technologies score highly by the Technology Readiness Level (TRL)[4] index, which is used as an indication of commercial readiness. A score of 6 and above is termed a "bankable asset" and only fusion technology scores below this (Figure 31).

Figure 31: Nuclear Technology Readiness Level

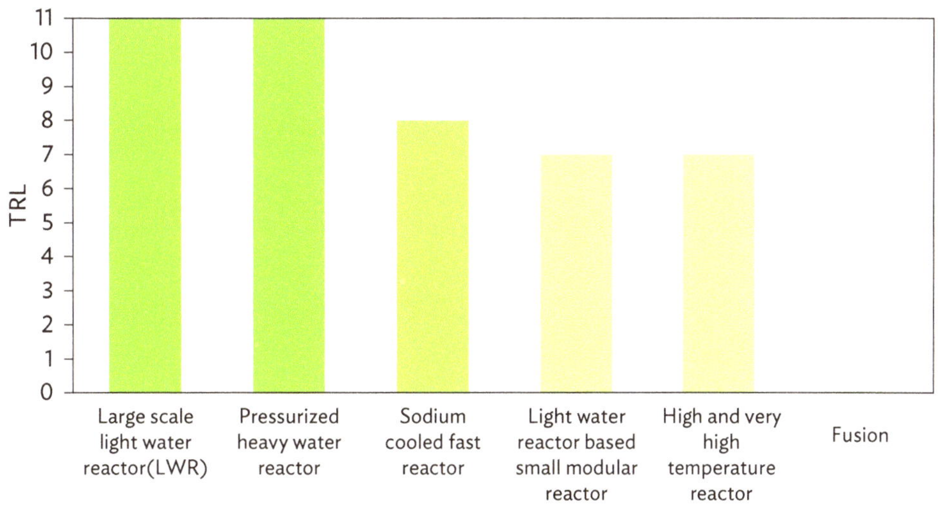

TRL = Technology Readiness Level.

Note: Commercial Readiness Index is as follows: (i) hypothetical commercial proposition (TRL 1-8); (ii) commercial trial, small scale (TRL above 8); (iii) commercial scale up; (iv) multiple commercial applications; (v) market competition driving widespread development; and (vi) bankable asset class.

Sources: Australian Renewable Energy Agency. 2014. Technology Readiness Levels for Renewable Energy Sectors.

The main design of reactors currently in use in Pakistan is the pressurized water reactor with five operational nuclear reactors. The government has a target to have 32 new nuclear power plants established by 2050. However, there are issues relating to the high capital investment and operating costs and safety concerns, coupled with the absence of a clear legal basis for managing civil nuclear liability, especially in the context of an accident with transboundary consequences. Not being a party to the Nuclear Non-Proliferation Treaty is also a barrier for technology development, although Pakistan is a member of the International Atomic Energy Agency. The government needs to prioritize using reactors with high safety levels and simplified systems. However, despite the long lead time and high capital cost involved, the technology offers a way of expanding generation capacity on a large scale.

Currently the most advanced technologies available in the sector are as follows:

- Gas-cooled fast reactor
- Lead-cooled fast reactor
- Molten salt reactor
- Sodium-cooled fast reactor
- Supercritical water-cooled reactor
- Very high-temperature gas reactor

[4] IEA. 2020. *Energy Technology Perspectives 2020*.

Table 11 summarizes the barriers and enablers facing the sector if the low-carbon scenario is to be implemented.

Table 11: Barriers and Enablers for Future Development of Nuclear Power

Parameter	Enablers	Barriers
Cost	• Reliable energy source.	• High capital costs.
Policy and Regulatory Environment	• According to the Pakistan Atomic Energy Commission (PAEC), the goal is to produce 8,800 megawatts (MW) electricity by 2030.[a] • Plans on constructing 32 nuclear power plants by 2050.[b] • New waste management offices are proposed by PAEC.	• Many licensing and regulation approvals are required. • Strict regulations on maintenance, staffing levels, operator training, and plant inspections. • Pakistan not a party to the Nuclear Non-Proliferation Treaty but is a member of the International Atomic Energy Agency.
Market	• Rapid economic growth means rising demand for electricity. • Relatively clean energy source. • Still commercial nuclear power plants generate only a small share of electricity.	• High operating costs. • Transportation, storage, and disposal of fuel is costly.
Technical and Commercial Performance	• Connected its first People's Republic of China-assisted 1,100 MW nuclear power plant in Karachi (K2) to the national grid in 2021.[c] • PAEC owns six plants, two in Karachi and four at Chashma in Mianwali district of Punjab.	• Long lead times and construction delays.
Stakeholder Engagement/ Perceptions	• PAEC has well-established communication with national and international stakeholders.	• Sometimes viewed by the public as a dangerous or unstable process.

[a] World Nuclear Association. Nuclear Power in Pakistan.

[b] *The Frontier Post*. 2021. Pakistan plans to generate 40,000 MW from nuclear power by 2050.

[c] *The Economic Times*. 2021. Pakistan connects its first 1,100 MW nuclear power plant to national grid. 19 March.

Source: Compiled by authors.

Hydro

Pakistan has vast untapped potential for hydropower, estimated to be 50–60 GW (footnote 1), with only 10,752 MW developed as of June 2022.[5] Around 27% of the total electricity demand in 2019 was provided by hydro-based electricity.[6] Under the low-carbon scenario, the share of hydro-based electricity generation would increase from 32% in 2025 to 38% in 2040 and then decrease to 32% by 2050 (Figure 32).

5 National Power Transmission and Despatch Co., Ltd. Indicative Generation Capacity Expansion Plans (IGCEP) 2022-31.

6 J. Unwin. 2019. What does Pakistan's energy mix look like and what is its future? *Power Technology*. 27 September.

Figure 32: Share of Electricity from Hydro
(%)

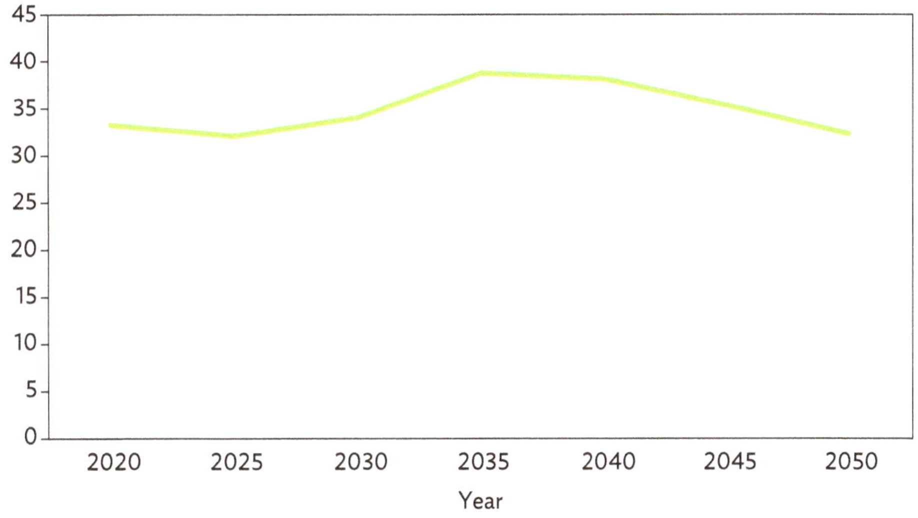

Source: Authors.

Hydropower generation is a mature technology. At the same time, development of a large-scale hydro project faces local challenges in relation to its impact on catchment areas and the resettlement of affected communities. These barriers, plus the high initial investment cost, are largely responsible for the slow development of hydropower projects in the country up to now. However, because of its cost-competitiveness, the optimization solution in the model shows a major expansion of hydro-capacity to nearly 60 GW by 2050 in the low-carbon scenario (Figure 33).

Figure 33: Hydropower Generation Capacity in a Low-Carbon Scenario
(GW)

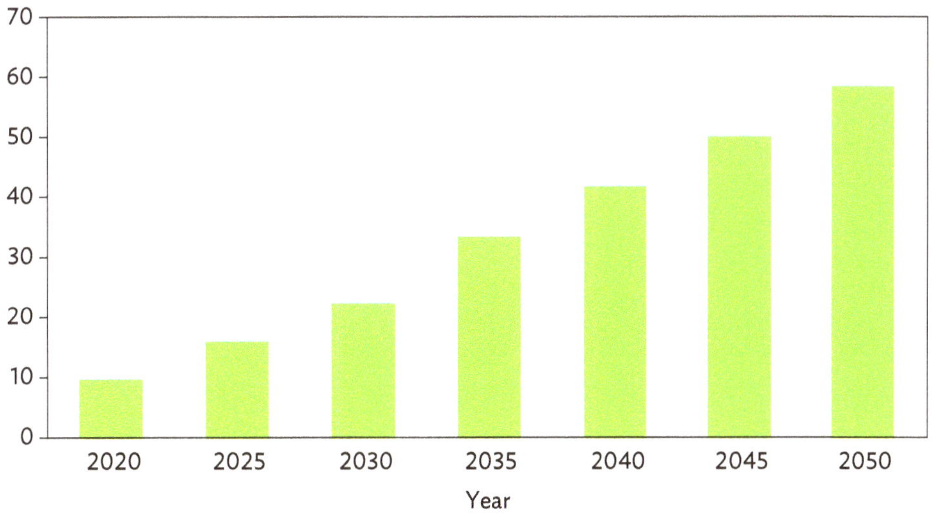

GW = gigawatt.
Source: Authors.

The two commercially applied technologies are large-scale and small-scale hydropower projects, usually distinguished by their capacity for power generation. Both are applied commercially and score 6 and 5 respectively on the TRL (Figure 34). It is particularly relevant when there is variable energy from renewable sources such as wind and solar. Innovative best practices in the application of the technology sector focus on developments in pumped hydro operations in combination with a variable renewable energy system to allow energy storage and grid balancing. So far, this is still not widely used in Pakistan but is an option for the future.

Figure 34: Technology Readiness Level for Hydropower Generation Technologies

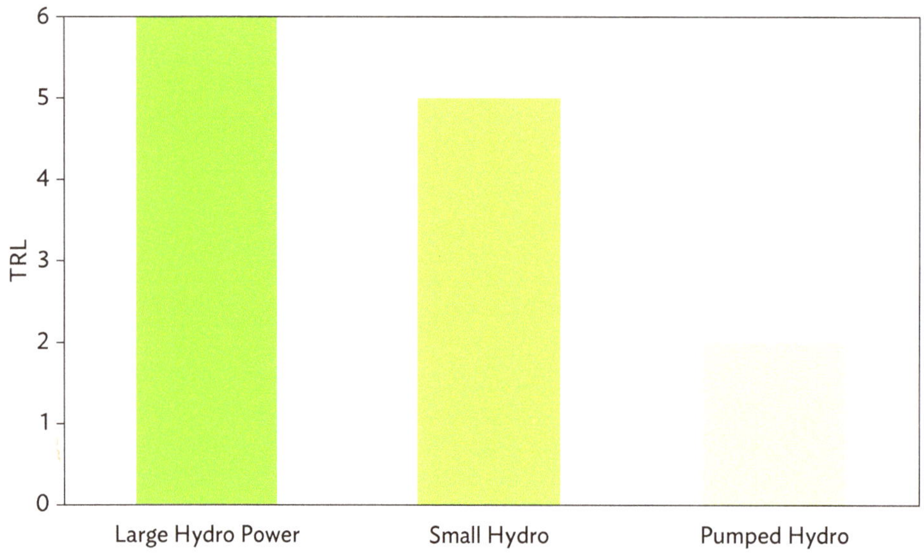

TRL = Technology Readiness Level.

Note: Commercial Readiness Index is as follows: (i) hypothetical commercial proposition (TRL 1-8); (ii) commercial trial, small scale (TRL above 8); (iii) commercial scale up; (iv) multiple commercial applications; (v) market competition driving widespread development; and (vi) bankable asset class.

Source: Australian Renewable Energy Agency. 2014. Technology Readiness Levels for Renewable Energy Sectors.

Hydropower development faces local challenges and needs to ensure that it resolves issues of community resettlement and protection of the environment and biodiversity. Nonetheless, it is one of the important technologies from the point of view of the cost of electricity and emissions reduction impact. Due to infrastructure requirements and often remote locations, the initial investment cost in hydro plants is high, but is compensated by the long life of such projects. In addition, growing renewables capacity will induce high variability into the electricity grid. This creates an opportunity for pumped hydro plants to protect against the variability of renewable sources and to provide daily and seasonal energy storage for the system.

Table 12 summarizes the barriers and enablers for hydropower if the low-carbon scenario is to be implemented.

Table 12: Barriers and Enablers for Future Development of Hydropower

Parameter	Enablers	Barriers
Cost	• Long economic life. • Low-cost electricity supply in long term.	• Very high capital investment requirement. • As the technology is mature, cost reduction potential is limited.
Policy and Regulatory Environment	• KP Hydropower Policy for Khyber-Pakhtunkhwa province 2016 mentions financial incentives and fiscal incentives. • New national electricity plan 2021 approved for the next 10 years. • Plan to reduce carbon footprints by increasing the share of renewable energy in power generation to 30% by 2030 from the current 4%.	• It has been argued that the National Electricity Policy 2021 has deprived the provinces of any role in the power sector. • Inclusion of the cost of transmission lines as an integral part of generation in draft policy makes the short-term cost of hydel power expensive.[a]
Market	• Potential of 60,000-megawatt (MW) hydroelectric power. Only 7,172 MW was generated from hydropower in 2016–2017. • High electricity demand expected in future. • 22 hydropower stations owned and operated by the Water and Power Development Authority (WAPDA) across the country generated 38.3 billion units. • Completion of ongoing projects by WAPDA can generate additional 11,122 MW.	• Lack of strategy and monitoring framework for hydro.
Technical and Commercial Performance	• Provides flexibility to the grid. Quick ramping up and down to meet peak demand and demand fluctuations. • Emerging pumped hydro storage and integration with variable renewable energy technologies. • Per capita energy consumption is increasing.	• Absence of adequate funding to improve small hydropower projects.[b] • Unavailability of data related to hydro resources, such as seasonal variation in the flow of rivers and streams, electricity demand, geology, and geography.
Stakeholder Engagement/ Perceptions	• Historically, hydropower has been the most important source of electricity and has wider acceptance. • Ongoing projects on Mohmand, Diamer Basha, Dasu Dam by WAPDA.	• Large hydropower projects face opposition due to environmental damage and loss of habitat of animals and vegetation. • Disagreement between the upstream and downstream villages on the sharing of water resources. • Displacement and rehabilitation challenges of affected people.
Skill Development Requirement	• As the technology is mature; skilled workforce is available for large hydropower projects.	• Small hydro and pumped storage projects are relatively new technologies. Workforce availability for design and development and grid integration of these technologies is limited.
End-of-Life	• Hydropower systems have a very long useful life; globally some have exceeded over a century. • Equipment used in hydropower generation can be replaced over time to extend its life.	

[a] R. Yusufzai. 2021. National Electricity Policy 2021: Draft policy termed anti-hydropower. *The News*. 21 June.

[b] W. Uddin et al. 2019. Current and future prospects of small hydro power in Pakistan: A survey. *Energy Strategy Reviews*, 24, pp. 166-177.

Source: Compiled by authors.

Natural Gas

Pakistan had 19 trillion cubic feet of proven gas reserves as of 2017, which is equivalent to 12 times its annual consumption.[7] Natural gas is cheap, abundant, and reliable, and produces 40% to 65% fewer emissions than coal.[8] Natural gas-based plants can be classified into the following:

- Combined cycle power plants
- Simple cycle gas turbine power plants
- Gas engines

All are widely used commercially proven technologies. With the global development of variable renewable energy, natural gas-based power has an important role to play in electricity grid balance and natural gas is considered one of the key fuels in the transition from coal-based generation.

The demand for natural gas will grow globally as countries seek to displace coal for heating and power generation to lower their GHG emissions. Historically, all natural gas production has been consumed domestically; however with high gas consumption and depleting domestic gas reserves, in the low-carbon scenario it is expected that the country will need to import liquefied natural gas (LNG) to meet the growing demand.[9] Hence, in the short term, investment in natural gas infrastructure is needed to handle the growing demand for imports of LNG. The Turkmenistan–Afghanistan–Pakistan–India natural gas pipeline, which has experienced a lengthy delay, could have played an important role if it were to be in place by now.[10]

In the low-carbon scenario, due to the depleting reserves the no new investment in gas-based plants is expected after 2030 and their share in electricity generation is estimated to decline from 23% in 2025 to 10% in 2040 and to disappear completely by 2045 (Figure 35). However, imports grow strongly reaching 40% of total energy imports and 26% of TPES by 2050. Total installed capacity reaches a peak in 2030 at just below 10 GW and then declines, being phased out by 2045 (Figure 36).

7 Worldometer, N.D. Pakistan Natural Gas. (accessed 22 June 2023).
8 Canadian Association of Petroleum Producers. World Energy Needs.
9 ANI. 2021. Pakistan to buy Liquefied Natural Gas cargo at highest ever rate: Report. 7 November.
10 ADB placed its regular assistance to Afghanistan on hold effective 15 August 2021. *ADB Statement on Afghanistan* dated 10 November 2021.

Figure 35: Share of Electricity from Natural Gas at a Low-Carbon Scenario (%)

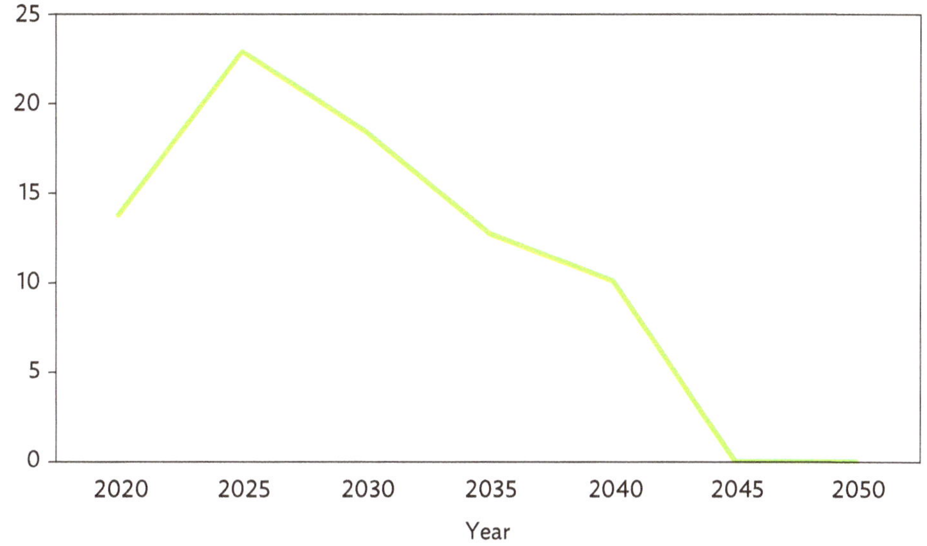

Source: Authors.

Figure 36: Gas-Based Power Generation Capacity Under a Low-Carbon Scenario (GW)

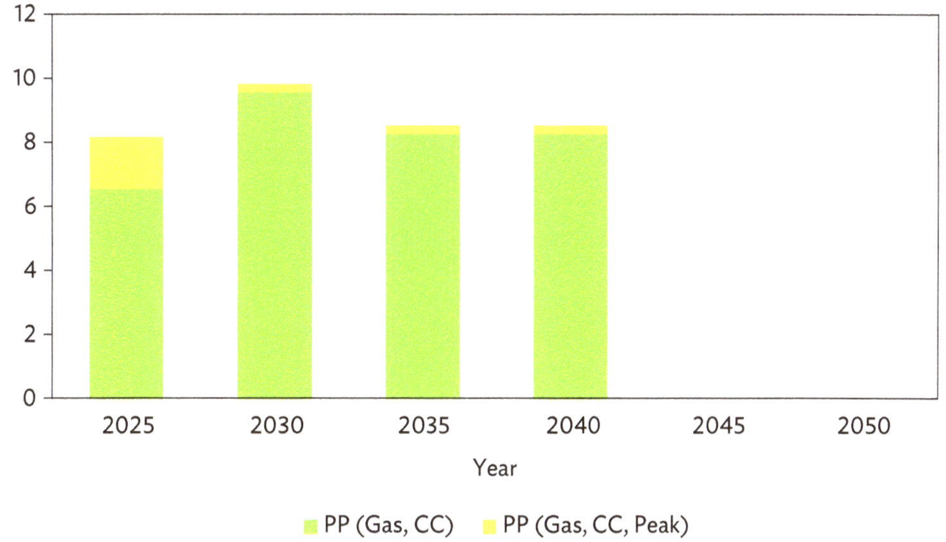

CC = combined cycle, GW = gigawatt, PP = power plant.

Source: Authors

Table 13 summarizes the barriers and enablers facing natural gas-based generation.

Table 13: Barriers and Enablers for Future Development of Natural Gas-Based Power

Parameter	Enablers	Barriers
Cost	• Low cost of power generation.	• High cost of infrastructure development especially liquefied natural gas (LNG) exploration and pipelines.
Policy and Regulatory Environment	• Incentives available in petroleum policy, 1997. • Pakistan signed a 15-year deal in 2016 to purchase natural gas from Qatar to supply the Engro Elengy terminal.	• Regulatory challenges, security concerns, and insufficient investment. • Lack of incentives for foreign exploration and production companies in upstream gas sector.
Market	• Infrastructure to import natural gas has improved after 2015 post commissioning of regasification terminals. • Rapid economic growth means rising demand for electricity.	• Decline of conventional natural gas reserves.[a] • Complex geography.
Technical and Commercial Performance	• Sizable shale gas reserves of 105 trillion cubic feet (United States Energy Information Administration's *Technically Recoverable Shale Oil and Shale Gas Resources Report*, 2013). • Natural gas contributes 38% of the country's total primary energy supply mix.[b]	• Technical expertise about exploration and supply of LNG is limited. • Insufficient investment. • Uncertainty in local production.
Stakeholder Engagement/ Perceptions	• Turkmenistan–Afghanistan–Pakistan–India (TAPI) natural gas pipeline was expected to commence operations in 2021.[c] • Availability of other sources of financing for development of natural gas pipeline infrastructure in the country.	• Security concerns. • Environmental constraints.
Skill Development	• Readily available, cost-effective labor force	• Lack of technology in exploration

[a] Government of the United States, US Energy Information Administration. 2016. Pakistan.

[b] International Trade Administration. Energy Resource Guide - Pakistan - Oil and Gas.

[c] Hydrocarbons Technology. Turkmenistan-Afghanistan-Pakistan-India (TAPI) Gas Pipeline Project.

Source: Compiled by authors.

Wind Power

The significant wind energy potential in the country provides one of the solutions for carbon-free energy generation. Over 9% of the geographic land area is suitable for most utility-scale wind turbine applications and total wind power generation potential is estimated to be over 349,000 MW.[11] The development of technology and increasing turbine size has led to significant price reductions. For example, the global levelized cost of electricity of a newly commissioned onshore wind energy

[11] S. Shami et al. 2016. Evaluating wind energy potential in Pakistan's three provinces, with proposal for integration into national power grid. *Renewable and Sustainable Energy Reviews* 53. pp. 408-421.

system has fallen from $0.089 per kilowatt-hour (kWh)[12] in 2010 to $0.039 per kWh in 2020, a 56% reduction. Similarly, in the case of an offshore energy system over the same period, the global weighted average levelized cost of electricity of offshore wind energy system has been reduced by 48% from $0.162/kWh to $0.084/kWh. Based on this potential and substantial cost reductions, under the low-carbon scenario, the share of wind-based electricity generation increases from 3% in 2025 to 14% in 2050 (Figure 37).

Figure 37: Share of Electricity from Wind-Based Electricity a Low-Carbon Scenario (%)

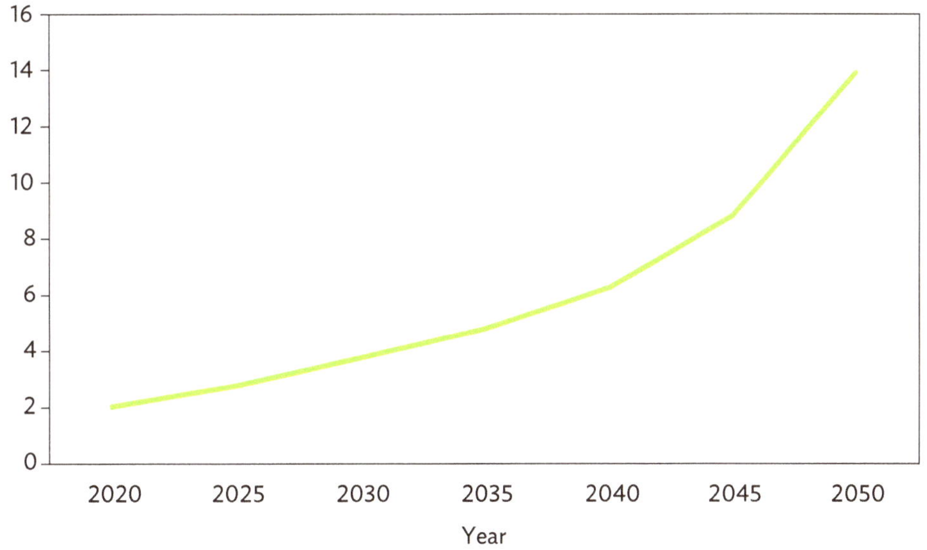

Source: Authors.

Wind power plants can be classified into three main categories:

- Onshore wind
- Offshore wind: Seabed fixed
- Offshore wind: Floating offshore

All three are now relatively mature and commercially applicable technologies with a TRL of 8 and above (Figure 38).

12 IRENA. 2021. Renewable Power Generation Costs in 2020.

Figure 38: Wind Technology Readiness Level

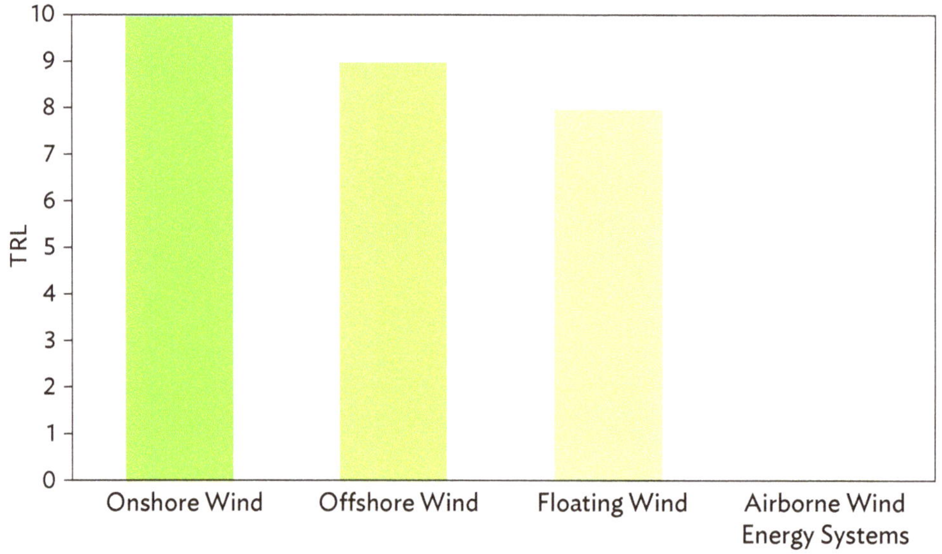

TRL = Technology Readiness Level.
Source: Authors.

Figure 39: Wind Power Installed Capacity Under a Low-Carbon Scenario (GW)

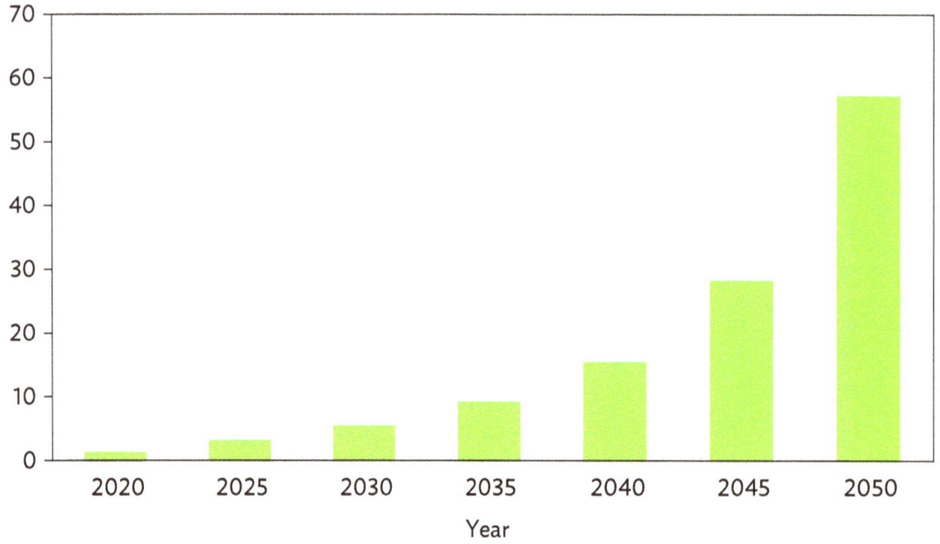

GW = gigawatt.
Source: Authors.

Wind-based power plays an important role in the country's long term generation expansion plan. There are policies and incentives in place such as the Alternative and Renewable Energy Policy 2019,[13] which aims to increase share of electricity generated from the renewable sources with wind and solar from 4% in 2019 to 20% by 2025 and 30% by 2030.[14] Rapid growth in wind energy-based power generation is expected and the Alternative Energy Development Board of Pakistan (AEDB) recently approved 11 separate wind-based projects with a collective capacity of 560 MW.[15]

Under the low-carbon scenario, installed capacity for wind power generation would grow from less than 5 GW in 2025 to 38 GW by 2050 (Figure 39). However, it should be noted that the limited domestic manufacturing capacity and construction capabilities for wind farms might act as barriers to this development.

Table 14 summarizes the barriers and enablers facing the sector if the low-carbon scenario is to be implemented.

Table 14: Barriers and Enablers for Future Development of Wind Power

Parameter	Enablers	Barriers
Cost	• Wind turbine prices have decreased by 44%–64% since 2010.	• Subsidized fossil fuel use.
Policy and Regulatory Environment	• Wind generators under 1 megawatt (MW) of capacity can sell back produced electricity to the national grid. • The Alternative and Renewable Energy Policy 2019 aims to increase share of electricity generated from renewable sources to 20% by 2025 and 30% by 2030. • Tax benefit to investors, incentives for local production of wind turbines.	• Limited budget allocation to the RE-based projects.
Market	• Rapid economic growth means rising demand for electricity. • Considerable wind potential available in different regions of Sindh and Balochistan provinces.[a]	• Lack of investment due to low competitiveness with global plants.
Technical and Commercial Performance	• Potential to generate 43,000 MW with a coverage area of 9,700 square kilometers with an average wind speed of 7.87 meters per second (m/s) in 10% of the windiest areas.[b] • Eight new wind farms with a combined capacity of 410 MW are expected to be operational by end of 2021.[c]	• Land availability.
Stakeholder Engagement/ Perceptions	• Increasing public awareness.	• Lack of information about credibility of developers.

continued on next page

[13] Government of Pakistan. 2019. *Power Policy*.

[14] International Finance Corporation. 2019. Wind Energy Transforms Pakistan.

[15] Mordor Intelligence. Wind Energy in Pakistan Market Size & Share Analysis—Growth Trends & Forecasts (2024–2029).

Table 14 *continued*

Parameter	Enablers	Barriers
Skill Development Requirement	• Low cost of labor and abundant labor supply.	• Unavailability of sufficiently trained engineers and technicians for manufacturing, design, construction, operation, and maintenance of wind power plants.
End-of-Life	• Wind turbines have long lives. During their operational lifetime the improvement in technology is significant. As a result, wind farms can be repowered with new wind turbines at the same location with high resource availability.	• There are limited standards to recycling of wind turbines. • The wind turbines made of composite materials are difficult to recycle.

[a] H.H. Zahid, Wei Jiang, and S. Rehman. 2017. Technical and Economic Assessment of Wind Power Potential of Nooriabad, Pakistan. *Energy, Sustainability, and Society.*

[b] Switzerland Global Enterprise. 2021. Market Snapshot: Renewable Energy in Pakistan. 17 August.

[c] T&D World. 2020. Pakistan to Ease Power Shortages with Eight New Wind Farms. 3 October.

Source: Compiled by authors.

Solar

Pakistan has immense potential to generate power through solar photovoltaic (PV). Barring a few areas in the northwestern part of the country, almost all areas receive abundant sunlight and have a dry and hot climate. The daily average solar power potential is 5.3 kWh/square meter (m^2).[16] Under the various types of installation, the key and most expensive component of the system is the solar PV panel. In the last decade, the global levelized cost of solar electricity of a newly commissioned system has fallen from $0.381[17] per kWh[18] in 2010 to $0.057 per kWh in 2020, an 85% reduction. The key drivers for the cost reduction were increased operating efficiency and reduced solar cell prices, which fell by 93% from 2010 levels.[19] While the capacity of the solar PV system has increased, due to its low utilization factor, its share in overall electricity supply will always be well below its share in capacity.

Power generation from solar energy can divided into the following categories based on the type of installation and connection to the grid:

- Ground-mounted large-scale solar PV
- Rooftop solar PV
- Floating solar power
- Solar PV with battery storage
- Concentrated PV

The technology is now relatively mature and can be applied commercially as shown by TRL estimates, with all but one having a TRL of 6 or above (Figure 40).[20]

16 M. Iftikhar et al. 2015. Sustainable energy for all in South Asia potential, challenges, and solutions. *Working Paper 151* (No. id: 12275).

17 As of 2020.

18 IRENA. 2021. Renewable Power Generation Costs in 2020.

19 E. Bellini. 2020. Pakistani regulator expects solar capacity to hit 27 GW by 2047. *PV Magazine.* 1 May.

20 IEA. 2020. Energy Technology Perspectives 2020.

Figure 40: Solar Technologies Readiness Level

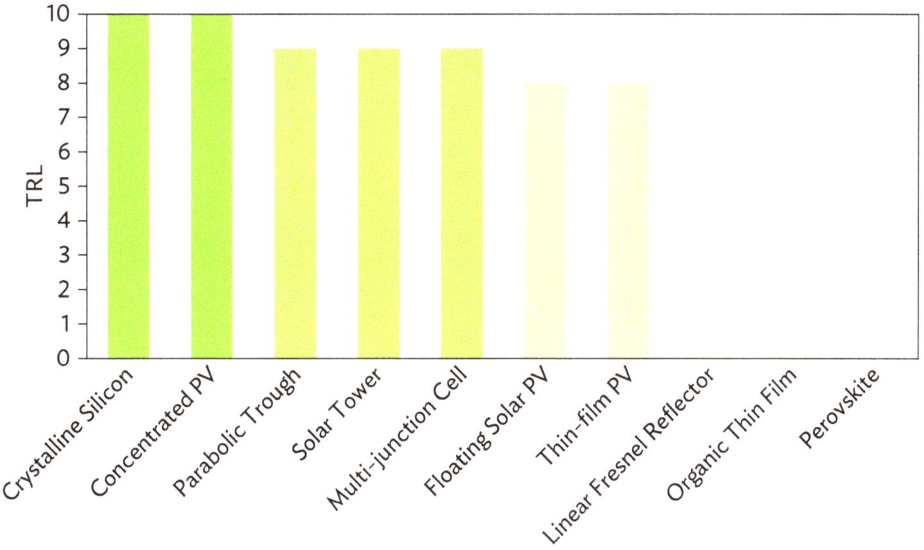

PV = photovoltaic, TRL = Technology Readiness Level.

Source: Authors.

Under the low-carbon scenario, solar PV power generation grows from 2% of total electricity generation in 2025 to 15% 2050 and installed capacity of solar PV from 5 GW in 2025 and to over 70 GW by 2050 (Figure 41 and Figure 42), which is 31% of total installed generation capacity.

Figure 41: Share of Electricity Generated from Solar Under a Low-Carbon Scenario (%)

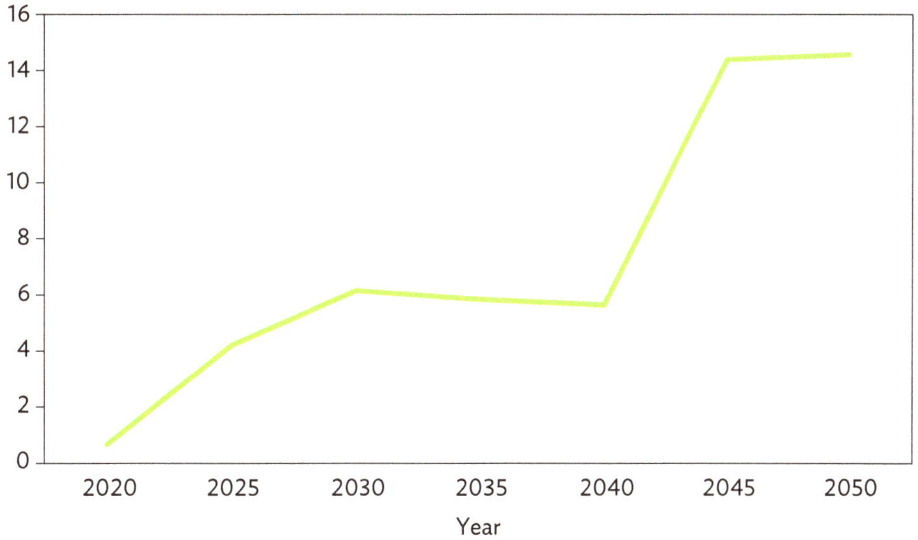

Source: Authors.

Figure 42: Solar Power Installed Capacity Under a Low-Carbon Scenario (GW)

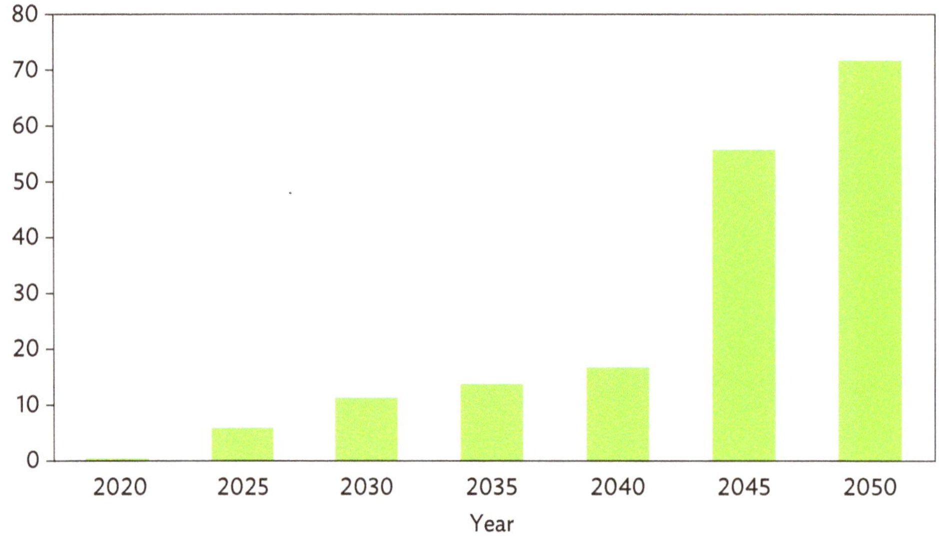

GW = gigawatt, PV = photovoltaic.
Source: Compiled by authors

From a very low base investment is now taking place in solar power. A solar energy project to power more than 200,000 homes in southern Sindh province was initiated in January 2021.[21] Financial incentives are available and provincial governments, particularly the government of Punjab, have taken the lead in facilitating development of solar power. However, more efforts are required if the country plans to achieve the low-carbon targets. Currently, all the components required to install solar panels need to be imported. The technology adoption rate can be improved by adding incentives for domestic solar panel manufacturing. Apart from the solar PV production and installation entities, it is also important to strengthen the electricity infrastructure to handle the variability of renewable energy. With a robust smart grid, it is necessary to have reliable grid availability for solar PV power producers.

Table 15 summarizes the barriers and enablers facing the solar power sector if the low-carbon scenario is to be implemented.

[21] S. Jamal. 2021. Pakistan solar project to power 200,000 homes in Sindh. *World Asia*. 29 January.

Table 15: Barriers and Enablers for Future Development of Solar Photovoltaic

Parameter	Enablers	Barriers
Cost	• Decreasing cost of solar installation	• Not as competitive in terms or pricing. • Inadequate market incentives.
Policy and Regulatory Environment	• Solar photovoltaic (PV) under 1 megawatt (MW) of capacity can sell back produced electricity to the national grid.[a] • Alternative and Renewable Energy Policy 2019 aims to increase share of electricity generated from renewable sources to 20% by 2025 and 30% by 2030. • Tax benefit to investors, incentives for local production of solar panels.[b]	• Limited follow up regarding implementation of policies.[c] • Limited budget allocation to the renewable energy-based projects.
Market	• Rapid economic growth means rising demand for electricity. • Currently produces only small share of electricity through solar power. • Installed PV capacity will likely increase from around 1.3 gigawatts (GW) at the end of 2019 to 12.8 GW by 2030 and 26.9 GW by 2047 (National Electric Power Regulatory Authority).[d]	• High interest rates on loans for procuring distributed energy systems. • Passive behavior of organizations responsible for solar PV technology uptake (footnote c).
Technical and Commercial Performance	• Geographic position gives great potential, Solar irradiance of 5.3 kilowatt-hour (kWh)/square meter (m^2)/day. • Solar power plant capital expenditure will fall from \$530/kW in 2020 to \$371/kW by 2030. • World Bank has provided \$100 million of financing to the Sindh Solar Energy Project.[e]	• Slow technological adoption. • Land shortage for harvesting solar energy. • Existing transmission infrastructure is old and there are huge losses.
Stakeholder Engagement/ Perceptions	• Acceptance of solar rooftop system.	• Lack of awareness and application of green building policies. • Lack of a better grid infrastructure with enhanced demand and capacity.
Skill Development Requirement	• Low cost of labor and abundant labor supply. • Training organized by Technology Upgrading and Skill Development Company and Pakistan Renewable Academy.	• Shortages of skilled workers in both front line and mid-level positions.
End-of-Life	• Solar panels have long life.	• Currently, no regulations exist for recycling. • After sales service is poor.

[a] International Energy Agency. 2015. Pakistan net metering policy for solar PV and wind projects. 4 September.
[b] M. Qasim. 2020. Pakistan leapfrogging to a green energy future. EastAsiaForum.
[c] F. Yazdan and M. N. Khan. 2018. Identifying and Addressing Barriers to Solar PV Technology Indigenization in Pakistan. *IJCSET*. 8(1). pp. 1–3.
[d] E. Bellini. 2020. Pakistani regulator expects solar capacity to hit 27 GW by 2047. *PV Magazine*. 1 May.
[e] Sindh Solar Energy Project. Pakistan, World Bank sign \$100 million agreement for 'Sindh Solar Energy Project.
Source: Compiled by authors.

Coal

In financial terms, coal is one of the cheapest sources of energy, which can be easily stored and transported. However, with the increased awareness of the effect of climate change and the damage from emissions, the government is planning to limit coal utilization and to focus only on the most efficient forms of coal-based electricity. Until 2016, the country only had one coal-based power plant, but currently, at least nine such power plants are under construction. Almost 19% of the power generation in the country came from just four coal-fired China–Pakistan Economic Corridor power plants in 2019-2021 and currently, coal accounts for about a fifth of total power output.[22] In 2021, the government expressed its intention of phasing out coal in the coming years at the Climate Ambition Summit.

Coal power plants can be classified into following categories:

- Subcritical coal power plant
- Supercritical coal power plant
- Ultra-supercritical coal power plant
- Integrated gasification combined cycle power plant.

Figure 43 shows their respective efficiency levels for a standard plant with subcritical plants the least efficient, and therefore most polluting.

Figure 43: Electricity Generation Efficiency of Coal Power Plants

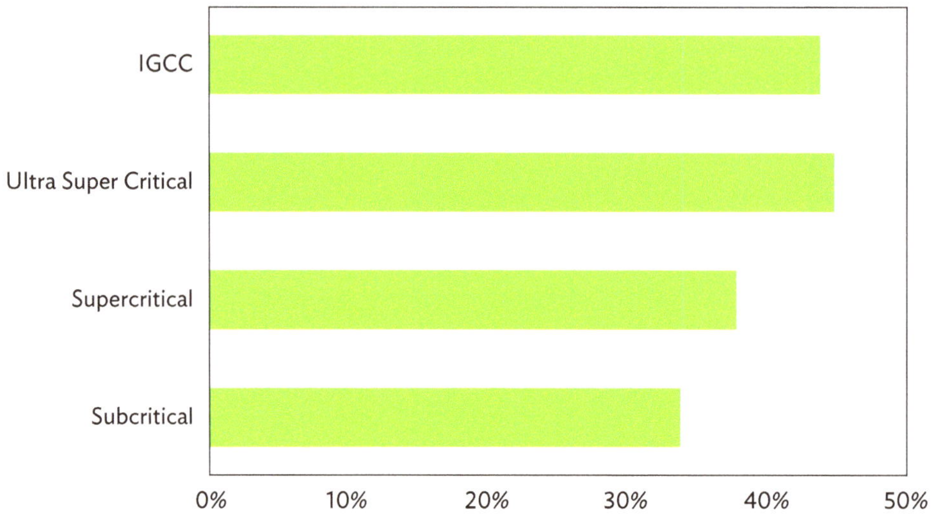

IGCC = Integrated gasification combined cycle power plant.
Sources: Global Energy Monitor Wiki. Coal power technologies; Energy Technology Systems Analysis Programme. Coal-Fired Power.

22 H. Isaad. 2021. Opinion: Is Pakistan really phasing out coal? *The Third Pole*. 18 January.

Although installed capacity of coal is expected to decrease after 2030, given depleting gas reserves and easy accessibility of coal reserves, some relatively efficient coal-based power is still expected to be needed. The reason for continued requirement of coal for electricity generation is the lack of alternative technology to meet electricity demand. So, even though the technology cost of renewables would be decreasing, the only alternative for meeting growing electricity demand would be coal-based power plants. This coal-based power generation will be expected to come from the more efficient coal supercritical, ultra-supercritical, and integrated gasification combined cycle plants.

Based on the low-carbon scenario, the coal power sector would reach a capacity of 9.3 GW by 2030, which decreases to 7.8 GW by 2050. The share of coal-based power in total generation would decline from 21% in 2025 to 13% in 2040, and 8% by 2050 (Figure 44). Coal supercritical technology-based power plants, which have the lowest emissions, would produce the entire coal-based electricity output by then.

Figure 44: Share of Coal-Based Electricity under a Low-Carbon Scenario (%)

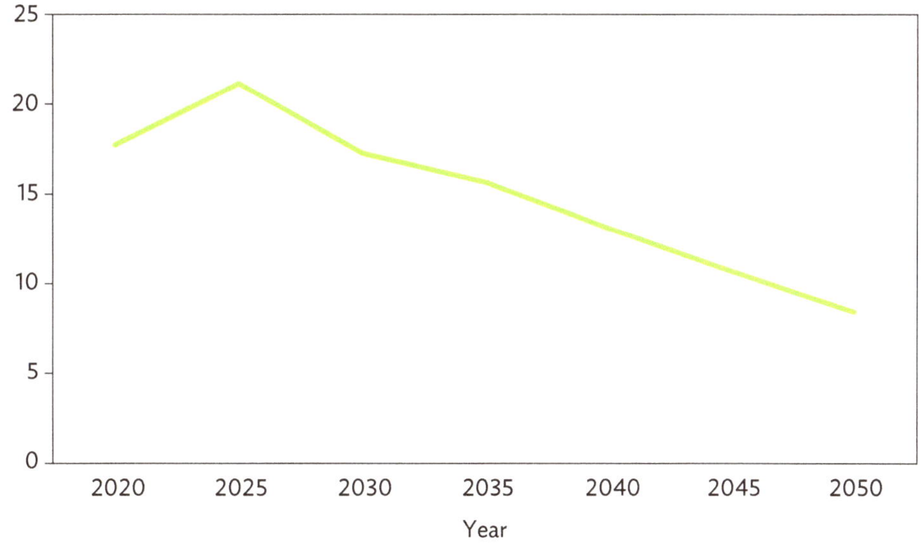

Source: Authors.

Table 16 summarizes the barriers and enablers facing the coal sector if the low-carbon scenario is to be implemented.

Table 16: Barriers and Enablers for Future Development for Coal-Based Power

Parameter	Enablers	Barriers
Cost	• Low cost of generation.	• Cost of the competing energy sources such as wind, solar, and natural gas is decreasing. • Increasing coal prices.
Policy and Regulatory Environment	• Pre-authorized seven new China–Pakistan Economic Corridor (CPEC) coal-fired power plants and fossil fuel projects will be implemented by 2024.[a]	• Alternative and Renewable Energy Policy 2019 aims to increase share of electricity generated from renewable sources to 20% by 2025 and 30% by 2030. • The Prime Minister of Pakistan announced a moratorium on coal at Climate Ambition Summit.
Market	• Coal-fired CPEC plants accounted for 19% of the country's power distribution in 2019–2020.[b] • Official data show Pakistan's coal-based power generation surged to 57% through fiscal year 2020.[c] • Coal power plants offer flexibility and reliability for the consumers.	• The government is likely to pivot from coal to imported liquefied natural gas for power production.
Technical and Commercial Performance	• 8.22 gigawatt (GW) capacity is coal-fired in nature in CPEC. • Solid investment that could deliver stable returns from operating plants. • Coal-to-liquid and coal-to-gas option plans are being explored to support growth in energy demand.	• Not all plants have been designed to provide flexibility under increasing share of renewables.
Stakeholder Engagement/ Perceptions	• Projects and funding support from the People's Republic of China under CPEC energy projects.	• Concerns against coal-based power plant by climate activists.
Skill Development	• Subcritical coal power plant is a mature technology and sufficiently skilled workforce is available in the country.	• The knowledge and training for new technologies such as carbon capture and supercritical coal thermal power plants is limited.

[a] R.S. Khan. 2020. Pakistan pushes renewables - but coal expansion continues too. *Reuters*. 8 August.

[b] H. Isaad. 2021. Opinion: Is Pakistan really phasing out coal? *The Third Pole*. 18 January.

[c] A. Gul. 2020. Pakistan Decides Against New Coal-fired Power. *Voice of America*. 12 December.

Source: Compiled by authors.

Supply-Side Emissions Mitigation

In the model, total mitigation is calculated relative to the emissions that would be created by the most polluting technology, taken as coal subcritical, and annual figures for emissions mitigated are summed. Similarly marginal abatement cost is calculated relative to the cost of abatement of the most polluting option. For example, for solar PV it is calculated from the following expressions:

$$Cost\_abatement_{SolarPV,year} = Levelized\ Cost_{SolarPV,year} - Levelized\ Cost_{Coalsubcritical,\ year} \quad \text{Equation 1}$$

$$Abatement_{SolarPV,year} = (Emission\ Factor_{Coal(subcritical),\ year} - Emission\ Factor_{SolarPV,year}) * Activity_{SolarPV,year} \quad \text{Equation 2}$$

$$Marginal\ Abatement_{CostSolarPV,year} (\$\ per\ tonne\ CO2\ abatement) = \frac{Cost\_abatement_{SolarPV,year}}{Abatement_{SolarPV,year}} \quad \text{Equation 3}$$

Where, Levelized cost covers all investment and operating cost discounted to a base year,[23] refer to the energy generated or consumed by a particular technology.

Table 17 summarizes the potential GHG mitigation effect over 2020–2050 of applying the investment needed under the low-carbon scenario. It also shows the estimated marginal abatement cost per ton of CO_2 equivalent for the different fuel sources.

Table 17: Supply-Side Mitigation Potential and Abatement Cost Different Fuels, Low-Carbon Scenario

Fuel Source	Mitigation Effect 2020–2050 ($MtCO_2e$)	Marginal Abatement Cost[a] ($\$/tCO_2e$)
Nuclear	674.0	10.2
Hydro	969.9	30.4
Natural Gas	138.6	3.6
Wind	225.9	5.7
Solar	322.8	(13.5)
Coal	60.9	30.9

() = negative, $MtCO_2e$ = million tons of carbon dioxide equivalent, tCO_2e = tons of carbon dioxide equivalent.

[a] 2022 prices.

Source: Compiled by authors

Hydro and nuclear offer by far the largest emissions impact, because of the scale of their investment under the low-carbon scenario. However, they are relatively high-cost fuels for abatement, given their capital intensity. The negative figure for the marginal abatement cost of solar reflects the fact that its levelized cost is less than that of the comparator coal subcritical and hence the numerator of equation 3 is negative. It does not mean solar energy is a costless means of abatement, only that its cost is below the coal comparator used. Solar, wind, and natural are the most cost-effective technologies for abatement, while supercritical coal and hydro are the least cost-effective.

Demand-Side Effects

These are also estimated in the low-carbon scenario for the five main user sectors. All the demand-side changes and technologies discussed in the following are commercially ready.

[23] Assuming a discount rate of 10%.

Buildings

The sector covers residential and commercial buildings. Apart from cooking, which the model examines separately, energy consumption in commercial and residential buildings, is in the following:

- Lighting
- Thermal comfort: heating, ventilation, and air-conditioning (HVAC) system, fans, air coolers, heaters
- Water heating: electric geysers, rods, and gas heater
- Electrical appliances: refrigerators, kitchen appliances, entertainment systems, computers, washing machines, and dish washers

In lighting efficiency improvements can be expected from the more widespread adoption of LED lights, which consume much less power than conventional bulbs. The energy consumed in lighting applications can also be significantly reduced using intelligent building automation systems, which turn lights on and off automatically based on occupancy using various sensors. Similarly, in HVAC systems, energy performance can be substantially improved with the use of building automation systems. For water heating, use of greener technologies such as rooftop solar water heating and electric heat pumps can significantly reduce energy consumption. For electrical appliances, improved labeling can help raise consumer awareness. Low-carbon development in the buildings sector will also require changes in building design such as thermal insulation and use of sustainable construction materials. Building integrated PVs is also emerging as a low-carbon alternative, which could have the dual purpose of window shading as well as power generation.

Under the low-carbon scenario, it is assumed that the obstacles to the application of these improvements can be overcome and there is a reduction of 8% in final energy demand from the buildings sector by 2050, as compared to BAU scenario, because of the application of energy efficiency measures.

Table 18 summarizes the barriers and enablers to the application of the energy efficiency measures needed if this is to be achieved.

Table 18: Barriers and Enablers for Energy Efficiency Application in Buildings

Parameter	Enablers	Barriers
Cost	• Monetary savings due to energy conservation.	• High upfront cost of investment. • The benefits are not directly visible.
Policy and Regulatory Environment	• Building Energy Code of Pakistan (BECP) developed in 1990. • National Energy Efficiency and Conservation Authority established in 2016 to introduce and implement energy conservation policies and practices. • "Free distribution of Compact Fluorescent Lamps (CFLs)" campaign to reduce household electricity consumption. • The Pakistan Energy Label for the Fan Industry provides certification for energy-saving fans.	• Need to revise and update the BECP energy code to cater to recent developments. • Lack of awareness regarding policy and incentives. • Lack of organizational and institutional setup regarding building energy efficiency in provinces other than Punjab.

continued on next page

Table 18 *continued*

Parameter	Enablers	Barriers
Market	• Building envelope has 40% energy-saving potential, high-efficiency lighting (LEDs) has 72% of energy conservation potential.[a]	• Best practice technologies and materials for energy conservation in buildings are often not available in the country. • Housing finance is low in Pakistan, with a mortgage-finance-to-gross domestic product ratio of 0.25.
Technical and Commercial Performance	• Potential for reduced emissions from improved energy efficiency given the size of energy demand from the buildings sector.	• Benchmarking energy use requires experts and energy auditors. Owners view this as an additional cost and are skeptical about the potential energy and monetary savings.
Stakeholder Engagement/ Perceptions	• High-end consumers are likely to demand building automation and smart systems.	• Inadequate knowledge about building energy performance and certification among developers and architects. • Consumers lack knowledge about energy consumption of household appliances.
Skill Development Requirement	• Renewable Energy and Efficiency Project with GIZ REEE is supporting the Punjab Energy Efficiency and Conservation Agency to finalize energy conservation building codes.[b]	• Trained workers are required to implement energy efficiency measures. • Skills to operate energy simulation programs are not available.
End-of-Life	• Equipment such as LED lights have significantly longer life than conventional items.	• Effective disposal of materials is a challenge.

[a] A. Ahmed. 2014. Energy smart buildings: potential for conservation and efficiency of energy. *The Pakistan Development Review, pp. 371–380.*

[b] GIZ. Renewable Energy and Energy Efficiency (REEE) Project.

Source: Compiled by authors.

Industry

The industry sector has a huge potential for energy savings. At present, producers most commonly use coal and oil-fired boilers for steam or hot water production. These are either water-tube boilers or fire-tube boilers, which create emissions from industrial process heating. To mitigate their environmental impact, energy efficiency and technology improvement measures can be adopted. These measures include the following:

- Combustion control optimization
- Flue gas heat recovery
- Soot blower optimization
- Chemical absorption for post combustion carbon capture
- Oxy-fuel combustion
- Chemical looping combustion
- Physical absorption of CO_2 from gasification of coal to syngas in water–gas–shift reaction

Alternatively, low-carbon alternatives such as natural gas, biomass, or hydrogen can be used in these boilers in fuel switching, or the energy source can be switched to solar thermal energy or electricity-based heating systems. Natural gas and biogas can replace coal, as they primarily contain methane which leads to a higher calorific value and lower emissions for process heating. Gas-fired boilers have an added advantage of flue gas condensation for steam recovery, which can improve the overall efficiency of the operation. Similarly, biomass can be a substitute fuel in a coal-fired boiler with little modification. Other technologies such as hydrogen and carbon capture and storage are also emerging, which have a high potential to mitigate emissions related to industrial process heating. However, as these technologies are still in nascent stages of development in the country.

There are insufficient environmental protection measures currently in force and a lack of awareness of the importance of energy efficiency for industry. The low-carbon scenario estimates a reduction of 10% in final energy demand in the sector by 2050, as compared to the BAU scenario. This is primarily due to an assumed transition to cleaner and more efficient energy sources, electrification, and embracing energy efficiency practices.

Table 19 summarizes the barriers and enablers to the application of the energy efficiency measures needed for industry if the low-carbon scenario is to be implemented.

Table 19: Barriers and Enablers for Application of Energy Efficiency Measures for Industry

Parameter	Enablers	Barriers
Cost	• Monetary savings due to energy conservation.	• Replacing old equipment with new energy-efficient technology requires investment.
Policy and Regulatory Environment	• National Energy Efficiency and Conservation Authority of Pakistan announced the country's first mandatory energy efficiency policy for electric motors in December 2020.	• Suitable regulatory measures and incentive mechanisms are needed for energy savings in the sector. • Insufficient industry environmental protection measures.
Market	• Gap between demand and supply of power leading to breakdown in manufacturing operations. • Considerable amount of energy-saving potential has been identified. The value of energy conservation opportunities in Pakistan are estimated to be of $5 billion per year with a share of 25% for industry. • Use of old and/or poorly maintained machines and poor energy management in manufacturing gives considerable scope for re-equipping leading to energy savings.	• Technology may not be commercially available in the country. • Financing of new technologies is a challenge.
Technical and Commercial Performance	• About 35% of national primary energy is consumed in industrial sector, so there is a huge potential for energy savings.[b]	• Huge investment and time are required for implementation. • Access to capital to improve energy efficiency is limited.
Stakeholder Engagement/ Perceptions	• Increasing awareness among industries about their role in sustainable development.	• Delay in response and insufficient resource allocation. • No dedicated energy manager in most small and medium enterprises. • Limited public budget for energy efficiency programs. • Limited or poor information regarding energy efficiency opportunities.
Skill Development Requirement	• A series of training courses for engineers has been conducted successfully.	• Workforce training is required to switch to sustainable methods. • As some of the technologies are still not widely used, a specialized workforce is needed.

[a] I. Ahmad et al. 2020. Drivers and barriers for efficient energy management practices in energy-intensive industries: a case-study of iron and steel sector. *Sustainability*. 12(18). p. 7703.

[b] M. Hassan, S. Burek, and M. Asif. 2017. Barriers to Industrial Energy Efficiency Improvement – Manufacturing SMEs of Pakistan. *Energy Procedia*. 113. pp. 135–142.

Source: Compiled by authors.

Transport

With a growing economy, it is important to decarbonize the transport sector to limit emissions. The sector largely consumes oil-based energy and there is significant scope for emissions mitigation. In addition, the high sulfur content in the fuels used for vehicles, a level of 5000–10,000 parts per million, is a concern.[24] This is significantly higher than Euro III or Euro IV emissions standards. Considering the global trend toward clean transportation modes, and assessing energy resource availability within the country, electrification of the transportation system is the most obvious technological option to mitigate emissions from the sector.

The electrification of transport requires the extensive use of electric vehicles either as battery electric vehicles or as plug-in hybrid electric vehicles. For their widespread use, the installation of a large-scale charging infrastructure, either domestic or public, is critical. Globally, there are charging standards adopted such as combined charging system Type 1 and Type 2 in the European Union and North America, GB/T in the People's Republic of China, CHAdeMO in Japan, and Tesla's own supercharging system. Each of these systems offers different functionality, charging loads, and communication protocols. The slow charging systems for electric cars range from 3 to 7 kW and can run on AC power. These are usually installed for domestic charging. The fast-charging systems have power ratings ranging from 50 kW to 400 kW of DC power and are installed at public charging stations.

The road transport emissions in the freight sector can be improved by upgrading emissions norms and deploying proper testing measures. Many countries mostly pattern their emissions policies on European regulations and the associated mandates for clean, low-sulfur fuels. By adopting the Euro 6/VI vehicle emissions standards, the country can achieve up to a 99%[25] reduction in the emissions of pollutants, like fine particulate matter (PM2.5) with associated health benefits.

At present, the penetration of electric vehicle in Pakistan is limited and the market is dominated by oil-based vehicles. There are significant barriers to transitioning to electric vehicles, such as limited charging infrastructure and power outages. Electricity breakdowns are a frequent occurrence, and hence electric vehicles may not work very successfully in the short term (footnote 23). However, there have been policy initiatives to support an ambitious National Electric Vehicles Policy and there has been private sector investment in the sector and the formation of the Pakistan Electric Vehicles Manufacturing Association (Table 20). The low-carbon scenario assumes that these barriers can be overcome and suggests a major fall of 21% in TFED from transport by 2050 relative to the BAU scenario.

[24] M.T. Sohail et al. 2021. Pakistan management of green transportation and environmental pollution: a nonlinear ARDL analysis. *Environ Sci Pollut Res.* 28. pp. 29046–29055.

[25] M. Williams and R. Minjares. 2016. *A technical summary of Euro 6/VI vehicle emission standards.* The International Council on Clean Transportation.

Table 20 summarizes the barriers and enablers to the introduction of electric vehicles.

Table 20: Barriers and Enablers for Electric Vehicles

Parameter	Enablers	Barriers
Cost	• Price range of e-motorbike is close to that for standard combustion motorbikes. • Cost of vehicles can be brought down by locally manufacturing and assembling vehicles.	• High upfront costs for e-cars.
Policy and Regulatory Environment	• National Electric Vehicles Policy[a] aims for electric vehicles to have 30% of all the passenger vehicle and heavy-duty truck sales by 2030, and 90% by 2040. • For two- and three-wheelers and buses; 50% of new sales by 2030 and 90% by 2040. Sales tax of 1% for electric as opposed to 17% for regular vehicles. • Import duty for charging equipment is 1%.	• Limited supporting infrastructure.
Market	• High demand is expected in near future. • Positive global trend.	• Nascent market. • Model availability. • Unstable electric power, frequent blackouts.
Technical and Commercial Performance	• Various power projects in the pipeline and forecast of surplus capacity generation, so electric vehicles can utilize that surplus (footnote a). • Manufacturing facilities of electric cars are planned, MG already established in Lahore.[b]	• Low availability of e-motorbikes. • High-load commercial usage is not feasible in the current number of electric vehicle charging points.
Stakeholder Engagement/ Perceptions	• Pakistan Electric Vehicles Manufacturing Association consisting of five domestic manufacturers have been making significant investments in the sector.	• Lack of awareness regarding how electric vehicles work. • Local market is not sizable enough to attract investments.
Skill Development Requirement	• Inexpensive labor.	• Challenge to attract manufacturers to develop electric vehicles domestically.

[a] M. Uddin. 2020. Pakistan's National Electric Vehicle Policy: Charging Towards The Future. The International Council on Clean Transportation. 10 January.

[b] M. Rizvi. 2024. Pakistan prepares to drive into the future with EV production. *Khaleej Times*. 24 January.

Source: Compiled by authors.

Cooking

The access to clean energy resources for cooking is limited in rural areas. Cooking consumes about 44% of the total energy in the country. In rural areas cooking is dominated by biomass, such as firewood and crop residues. The major source of fuel for urban cooking is liquefied petroleum gas, with a small proportion from dry biomass. Increased use of gas within the overall cooking mix, especially in urban areas, has led to a reduction in energy demand for cooking. Currently, electric-based cooking is in a nascent stage, as the lack of reliability of power supply has a strong negative effect on its acceptance. There is an opportunity to use roof top solar, incorporated with electric cooking, to get more than 50% of the cooking energy from the roof top PV, but progress is slow, and more initiatives and policies are required to boost the acceptance of this mode of cooking.

The low-carbon scenario assumes the barriers discussed in Table 21 can be overcome and shows a 10% reduction in final demand in the sector in 2050 relative to the BAU scenario.

Table 21 summarizes the barriers and enablers to the introduction of electric cooking.

Table 21: Barriers and Enablers for Energy Efficiency in Cooking

Parameter	Enablers	Barriers
Cost	• Operating costs of induction stoves are lower than electric filament stoves and liquefied petroleum gas (LPG) stoves. • Biomass is cheap and readily available in rural areas.	• Innovative business and financing models needed to reduce distribution costs and increase affordability. • High upfront cost for biogas plant.
Policy and Regulatory Environment	• Zero advance tax on LPG imports.	• More incentives needed for the private sector to invest in clean energy fuel for cooking. • Government's policies favor imports over local production of LPG, as there is a tax disparity between local LPG and imports.[a]
Market	• 68% of households in rural areas use fuelwood, 15% use natural gas, and 17% use crop residues for cooking.[b]	• Affordability issues. • Lack of reliability of power supply affects acceptance rate in electric cooking.
Technical and Commercial Performance	• Cooking consumes about 44% of total energy. • Annual biomass consumption is equivalent to around 17 million tons of oil. • An approximate investment of PRs2.38 billion was made in the LPG supply infrastructure in 2016.[c]	• Lack of access to alternative technologies. • Lack of affordability of cleaner fuels. • Access to short-term finance for capital investment, developing distribution chains, and working capital.
Stakeholder Engagement/ Perceptions	• Opportunity to use roof top solar to be incorporated with electric cooking.	• Lack of awareness of health benefits from cleaner fuels and sociocultural reasons associated with fuelwood burning.
Skill Development Requirement	• Pakistan Poverty Alleviation Fund conducted community training regarding sustainable procedures and smokeless fuel-efficient cooking[d]	• School curriculum can be used to increase awareness around clean cooking. • Low awareness of the use of other technologies.

[a] *The Express Tribune*. 2021. Govt changes stance on LPG policy. 11 July.

[b] M. Saad Moeen et al. 2016.Factors Influencing Choice of Energy Sources in Rural Pakistan. *The Pakistan Development Review*. 55:4 Part II (Winter). pp. 905–920.

[c] Energypedia. Pakistan Energy Situation (accessed 15 January 2024).

[d] Pakistan Poverty Alleviation Fund. 2021. *Environment and Social Management Framework*.

Source: Compiled by authors.

Agriculture

Energy demand in the agriculture sector stem mainly from pumps and mechanization. Pumps create the energy consumption for irrigation and mechanization creates the energy consumption from land use. Currently, most farms use diesel-powered pumps for water irrigation, often running at suboptimal loading, which leads to higher energy consumption. The main low-carbon alternatives are as follows:

- Electrification of pumps
- Solar pumps

Electric pumps are equipped with electric motors, usually single or three phase induction motors, which drive the centrifugal pumps. They have higher operating efficiency as compared to diesel pumps and the newer electric pumps have a controller for pump speed and thus, the water flow rate. Also, they have water level sensors, which turnoff the pump when the water level reaches below a set limit, thus preventing a dry run of the pump.

Solar pumps are electric pumps connected to solar PV system. The PV modules convert the solar energy into electricity, which is used to drive the pumps. Their operating cost is low as the primary energy is virtually costless. Off-grid systems are stand-alone solar-powered pumps which operate during the availability of solar energy. The main advantage of these pumps is that they are not dependent on grid electricity. On-grid solar pumps can be operated when the sun is shining but also at night, in case of emergency. These systems offer the advantage of selling the excess electricity generated to the distribution company for additional revenue for farmers.

Despite some of the obstacles (discussed in Table 22), the low-carbon scenario is based on a rapid growth in both electric and solar pumps with diesel pumps completely replaced by 2050. Under the low-carbon scenario, off-grid electrical pumping would grow significantly, accounting for about 10% of energy demand by 2030 and reaching 21% by 2050, and final energy demand from agriculture is 26% below that in the BAU scenario by 2050.

Table 22 summarizes the barriers and enablers to the introduction of electricity-based pumping and mechanization.

Table 22: Barriers and Enablers for Future Development of Energy Efficiency in Agriculture

Parameter	Enablers	Barriers
Cost	• Operating cost and maintenance for solar pumps is negligible. • Help in reducing annual subsidy of more than PRs40 billion for life of solar pumps.[a]	• High initial costs of material and installation of solar pumps.
Policy and Regulatory Environment	• Government has set a target of 30,000 solar irrigation pumps installed on a budget of $93.2 million over 5 years. • In 2016, the provincial government of Punjab in collaboration with the Asian Development Bank, provided an 80% subsidy for installing solar irrigation pumps.[b] • Punjab Irrigated Agriculture Productivity Improvement Program Project seeks to stimulate the use of drip irrigation technology through capital cost subsidies.	• Lack of specific solar irrigation policy.
Market	• Groundwater utilization in Pakistan meets 60% of the total irrigation water requirements of the country (footnote b). • Huge potential of solar irrigation system.	• Installation, and operation and maintenance of solar pumps is difficult.

continued on next page

Table 22 *continued*

Parameter	Enablers	Barriers
Technical and Commercial Performance	• Most tube wells operate below rated capacities and require retrofits. • Higher energy efficiency of electric pumps. • Reduced emissions with solar pumps. • Solar pumps prevent overexploitation of water resources. • Effective control and demand-side management of power.	• Lack of funding for research and development to produce energy-efficient solar pumps locally.
Stakeholder Engagement/ Perceptions	• Farmers with diesel pumps can shift to electric and solar pumps due to lower operating cost	• Demonstration projects at the community level are required for wider acceptance of solar pumps.[c]
Skill Development Requirement	• Successful pilots undertaken by ENERCON and through the Tube-well Efficiency Improvement Program, supported by the United States Agency for International Development, have shown that significant improvement in tubewell efficiency is possible.[d]	• Lack of knowledge among farmers regarding best practices in agriculture.
End-of-Life	• Solar and electric pump systems have longer life than diesel pumps.	• Disposal of solar modules at the end of working life will require establishment of appropriate solar waste collection mechanisms.

[a] J. Ayaz. 2015. Potential of solar irrigation water pumps in Pakistan. In: Intersolar Conference. Munich, Germany, Conference Date: 10–12 Jun 2015.

[b] International Water Management Institute. Pakistan.

[c] M. Irfan et al. 2019. Solar Energy Development in Pakistan: Barriers and Policy Recommendations. *Sustainability*. 11. 10.3390/su11041206.

[d] World Bank Group. (n.d.) Energy Efficiency Roadmap for Pakistan.

Source: Compiled by authors.

Demand-Side Emissions Mitigation

The results from the low-carbon scenario in terms of emissions mitigation and the associated marginal abatement cost are summarized in Table 23.

Table 23: Demand-Side Mitigation Potential and Abatement Cost User Sectors in a Low-Carbon Scenario

Fuel source	Mitigation Effect 2020–2050 ($MtCO_2e$)	Marginal Abatement Cost ($/$tCO_2e$)
Residential and Commercial	1,782.6	(18.1)
Industry	1,510.7	(13.3)
Transport	1,221.7	(77.3)
Cooking	1,000.7	2.0
Agriculture	58.2	14.3

() = negative, $MtCO_2e$ = million tons of carbon dioxide equivalent, tCO_2e = tons of carbon dioxide equivalent.

Source: Compiled by authors.

As earlier discussed, the calculations are a comparison with the most polluting alternative and are thus illustrative of a hypothetical scenario rather than actual predictions. Nonetheless, they show great potential for an impact on the demand side through improved practices and technical improvements. Most of the mitigation effect arise from three areas: residential and commercial, industry, and transport. Emissions reduction in transport, in particular, is potentially both large and cost-effective. Again, negative marginal abatement costs arise because the technologies involved are compared with the most polluting coal alternative and where their cost is below that of the coal alternative the marginal figure is negative.

CHAPTER 4

INVESTMENT COST AND FINANCING OPTIONS

The chapter identifies the investment required for different sectors and power generation technologies derived from the model. These numbers depict the incremental cost for the capacity addition over the already existing capacity within the system, in 2022 prices. It then discusses the funding sources available to support these investments.

Investment requirements up to 2050 will be large in both scenarios but are higher in the low-carbon scenario. Table 24 gives totals over different sub-periods and Tables 25 and 26 give the breakdown by technology and user sector for the BAU and low-carbon scenarios, respectively.

Table 24: Investment Required Under Both Scenarios
($ billion)

	2020–2030	2030–2040	2040–2050
Total Investment (Business-as-Usual Scenario)[a]	125.0	176.2	292.8
	2020–2030	**2030–2040**	**2040–2050**
Total Investment (Low-Carbon Scenario)[a]	157.0	227.9	401.1

[a] 2022 prices.
Source: Compiled by authors.

Tables 25 and 26 give the 5-year totals for investment (so, for example, the figure for 2025 is the total investment for sub-period 2020–2025). In Table 24, figures for the different periods are the sum of the relevant 5-year totals. Hence for 2030–2040, for example, total investment in the sub-period is the sum of the figures for 2035 and 2040 from Tables 25 and 26.

Under BAU, major investments would be required for nuclear, hydropower, and coal supercritical power plants. Within demand sectors, major investment would be in transport and industry, for electrification and the transition to clean energy sources.

Table 25: Investment Requirement Under a Business-as-Usual Scenario
($ billion)

Sectors	2025	2030	2035	2040	2045	2050
Agriculture	1.1	0.8	0.9	1.1	1.1	1.2
Commercial	0.6	1.4	1.4	2.6	3.0	4.7
Residential (Cooking)	0.8	1.0	1.1	1.4	1.7	1.9

continued on next page

Table 25 *continued*

Sectors	2025	2030	2035	2040	2045	2050
Industry	1.6	2.7	5.6	7.5	9.8	14.4
Residential (Other Appliances)	1.0	1.1	1.3	1.5	1.8	2.1
Transport (Freight)	24.9	12.8	34.8	27.7	52.2	53.0
Transport (Passenger)	4.2	5.9	6.3	7.8	8.4	10.5
Domestic (Resource)	0.0	0.0	0.1	0.6	0.1	0.1
Gas	0.2	0.3	0.4	0.6	0.4	0.6
Imports	0.0	0.1	0.4	0.4	0.8	2.3
Oil	0.1	0.2	0.2	0.3	0.4	0.5
Hydro Power	0.0	28.2	7.9	8.4	9.0	9.7
Onshore Wind	2.3	2.7	2.6	2.8	3.8	15.2
PP (Biomass)	0.1	0.0	0.5	0.4	0.3	0.3
PP (Coal, SC)	5.8	1.3	6.0	7.1	9.4	9.1
PP (Gas, CC)	1.1	3.5	0.0	1.6	1.9	0.2
PP (Nuclear)	3.4	3.3	10.5	11.6	17.6	26.8
Solar PV	4.8	4.2	4.0	3.6	3.6	6.0
Electricity Grid	1.6	2.0	2.4	3.0	3.8	5.1
Total	**53.5**	**71.5**	**86.4**	**89.8**	**129.2**	**163.6**

CC = combined cycle, PP = power plant, PV = photovoltaic, SC = super critical.

Note: Figures are 5-year totals.

Source: Compiled by authors.

Under the low-carbon scenario over the whole period, an additional investment above that in BAU of around $192 billion would be required if the low-carbon path is pursued. Compared with the BAU scenario, this is an increase of over 30% in constant prices. Major investment is required particularly in hydro, nuclear, solar, and wind technologies. Coal and natural gas-based power plants would not require any investments from 2030 onward. Most of the demand-side investments would be required in the transport, residential, and industrial sectors.

In the transport sector, low-carbon investments are required for introduction of electric vehicles, but the estimates are broadly similar to those in the BAU because the additional new demand for mobility resulting from socioeconomic growth supersedes the reduction in energy demand from electrification of transport and stabilizes the demand. In the industry sector, the investments under low-carbon scenario are lower than the BAU scenario resulting from efficiency improvements because of the change in system efficiency due to fuel shift and corresponding lower investments to meet the same level of demand.

The major difference is the much larger investment required in the domestic resource sector[26] under the low-carbon scenario on account of increased investment in domestic gas and off-grid solar for application in agriculture and residential sectors.

[26] Domestic resources include domestic solar off-grid, wind off-grid, coal, gas, oil and biomass.

Table 26: Investment Requirement Under a Low-Carbon Scenario
($ billion)

Sector/Technology	2025	2030	2035	2040	2045	2050
Agriculture	1.2	0.8	1.2	1.3	2.2	1.9
Commercial	0.9	1.5	1.8	2.8	3.7	5.4
Residential (Cooking)	0.9	1.1	1.4	1.6	2.0	2.5
Industry	1.4	2.3	3.8	6.9	9.8	12.6
Residential (Other appliances)	1.0	1.1	1.3	1.5	1.8	2.1
Transport (Freight)	25.0	12.9	35.1	28.0	52.5	53.3
Transport (Passenger)	4.3	5.7	6.2	7.2	8.2	9.8
Domestic (Resource)	1.0	1.6	2.9	6.0	9.8	18.3
Gas	0.2	0.5	0.6	0.6	1.0	1.3
Imports	0.0	0.3	0.6	0.7	1.4	2.2
Oil	0.0	0.1	0.2	0.2	0.3	0.4
Hydro Power	23.3	22.1	35.8	25.5	24.1	22.7
Onshore Wind	2.3	2.7	4.2	7.0	15.0	31.3
PP (Biomass)	0.0	0.4	1.0	0.4	0.4	0.8
PP (Coal, SC)	3.9	4.0	0.0	0.0	0.0	0.0
PP (Gas, CC)	2.2	3.5	0.0	0.0	0.0	0.0
PP (Nuclear)	5.3	9.2	13.3	18.3	25.1	31.4
Solar PV	5.0	4.4	1.9	2.1	25.0	12.4
Electricity Grid	2.5	2.3	2.9	3.4	4.7	5.7
Total	**80.4**	**76.6**	**114.3**	**113.6**	**187.1**	**214.0**

CC = combined cycle, PP = power plant, PV = photovoltaic, SC = supercritical.
Note: Figures are 5-year totals.
Source: Compiled by authors.

Under both scenarios, substantial investment is needed, and the following sections discuss the potential sources of funds, particularly over the medium term (up to 10 years).

Government Initiatives

Investment in renewables and improved energy efficiency are now a policy priority.[27] As part of this drive the Alternative Energy Development Board (AEDB)[28] was created by the Ministry of Energy Power Division to facilitate, promote, and encourage development of renewables. AEDB is mandated to encourage the transfer of technology and to develop an indigenous manufacturing base for alternative energy technologies to meet the ambitious goal of 20% of energy capacity taken by renewable technology in 2020 and 30% in 2030 (Table 27).

27 Government of Pakistan, National Electric Power Regulatory Authority. 2020. *State of Industry Report 2020*.
28 AEDB is the sole representing agency of the Government of Pakistan.

Table 27: Initiatives to Promote Energy Sustainability in Pakistan

Initiative	Details
Pakistan Vision 2025 (2014)	The policy was issued by the Ministry of Planning Development and Reform. It aims to (i) increase power generation to over 45 gigawatts (GW), (ii) increase electricity access to 90% of the population by 2025, (iii) reduce the cost of generation by 25% by developing an optimal generation mix, (iv) reduce distribution losses by 10% (the average transmission and distribution loss in 2020 was around 21%),[a] and (v) undertake demand-side management measures by increasing use of energy-efficient appliances and products to 80%.
National Sustainable Development Strategy (2012)	The national road map for the implementation of Sustainable Development Goal 7 involves the following targets for 2030: (i) 100% access to electricity;[b] (ii) renewables 60% of total generation; and energy intensity of less than 2 megajoules/$, measured in terms of primary energy and gross domestic product.
Remote village electrification program	Alternative Energy Development Board (AEDB) is to provide electricity to 7,874 remote villages in Sindh and Balochistan provinces through renewable technologies by 2030.[c]
Solar PV power project	• This project is executed by AEDB with Zhenfa Pakistan New Energy Company (Private) Limited (ZPNECPL). Under this project, ZPNECPL will set up 229,885 solar photovoltaic modules of 435-watt peak in Layyah district, Punjab, each having a total capacity of 100 megawatts (MW). The modules will produce 188,427,600 kilowatt-hours (kWh) of clean electricity per annum. The levelized tariff for the project is $3.72/kWh.[d] • Additionally, as of March 2020, the National Electric Power Regulatory Authority has approved net-metering of more than 4,125 solar installations having a cumulative capacity of greater than 75 MW.[e]

[a] Government of Pakistan, Ministry of Planning, Development and Reform. 2013. *Pakistan Vision 2025*

[b] The Pakistan Credit Rating Agency Limited. 2021. Power Transmission & Distribution.

[c] T. Aized, S.M.S. Rehman, and M. Sumair. 2021. *Pakistan energy situation, policy, and issues. In Recent Advances in Renewable Energy Technologies*. pp. 387-428.

[d] F. Yousafzai. 2021. AEDB, ZPNECPL sign accord for 100mw solar PV power project. *The Nation*. 2 February.

[e] H. Waleed. 2020. AEDB assists SBP in revising financing schemes for renewable energy. *Business Recorder*. 19 July.

Source: Compiled by authors.

Private Sector Investment

Historically, the energy sector has been government-funded. Private participation in the power generation sector was first introduced in 1987. The HUB Power Company was the first private company to invest in a power plant with a capacity of 1,292 MW. Private sector participation in the sector picked up with the introduction of independent power producers in 1994, followed by the privatization of Karachi Electric Supply Corporation (now K-Electric), in 2005. The government initiated the privatization of generation companies in 2005 to address their budget deficits and improve profitability. Severe power shortages, expensive power generation, and lack of government funds to invest in the sector necessitated the involvement of private participants.[29] The National Electric Power Regulatory Authority Report for 2018 reported that private sector participation in installed generation capacity increased from 48% (11.4 GW) in 2014 to 57% (20.5 GW) in 2018, showing a substantial growth in investor confidence and a supportive policy environment.[30]

[29] World Bank. 2014. Pakistan: Powering Homes and Businesses. *Partnerships*. 2 January.

[30] National Electric Power Regulatory Authority. State of Industry Report 2018.

Transmission activity was liberalized with the introduction of the Transmission Line Policy in 2015. This envisages the development of public–private partnership projects under the build-own-operate-transfer model for 25 years, after which time ownership would be transferred to the National Transmission and Despatch Company.[31]

Continuing with the privatization of the generation sector, in November 2018 the Council of Common Interests moved to privatize two regasified liquefied natural gas (RLNG)-based power plants (owned and operated by the state-run National Power Parks Management Company Ltd) with a combined output of over 2,400 MW that accounts for 7% of the country's energy requirements.[32, 33]

Figure 45 gives the investment through private participation in infrastructure for electricity 2015–2020.

Figure 45: Private Participation in Infrastructure in Electricity Sector
($ million)

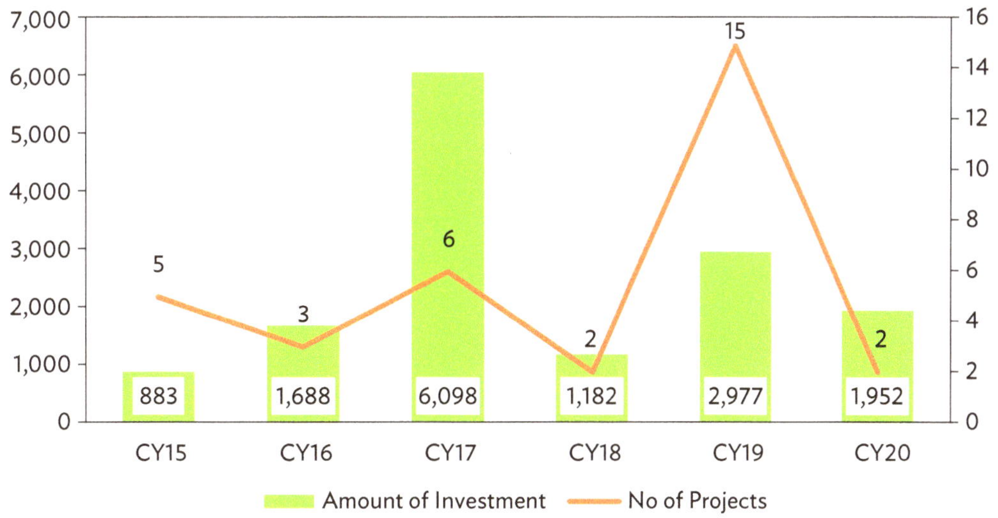

CY = calendar year, H1 = first half of a calendar year.
Source: World Bank. Country Snapshots (accessed 12 May 2020).

31 The policy would enhance the transmission network by developing reasonably sized financeable packages thereby comprising one or more transmission lines along with the associated grid infrastructure.

32 I. A. Kundi. 2019. Plan to complete privatisation of two power plants in FY 2019-20. *The Nation*. 30 June.

33 S. Iqbal. Dynamism in Pakistan's power sector: new avenues for private sector involvement in power infrastructure - China Working Group. International Bar Association; I. A. Kundi. 2019. Plan to complete privatisation of two power plants in FY 2019-20. *The Nation*. 30 June; and M. Haider. 2021. Circular debt delaying privatisation of two RLNG-based power plants. *The International News*. 1 November.

However, private investments in the energy sector are focused predominantly on fossil fuels such as coal and RLNG. The investments in coal-based power plants were driven by investments made under the China–Pakistan Economic Corridor. Further, the push toward coal and RLNG was to reduce the dependency on expensive oil-based power generation. With an ambition to increase substantially the share of renewable energy and to meet its international commitments on sustainability, the government needs to strengthen its regulations and policies to attract private and foreign capital into renewables. To increase the foreign direct investment inflow in this area, AEDB needs to review and strengthen the implementation of policies such as the Alternative and Renewable Energy Policy, 2019, and more generally the government needs to consider the wider investment climate for foreign investors.[34]

Compared to its regional competitors, Pakistan is still underdeveloped in terms of attracting Foreign direct investment primarily due to its unstable tax structure, lengthy dispute resolution process, and poor intellectual property rights enforcement. These weaknesses explain its relatively low ranking in the World Bank Ease of Doing Business index (108th out of 190 countries in 2020).[35] In 2019–2020 the power sector (including thermal, coal, and hydro) received investments of $764.3 million, or 29% of the total investments (Figure 46).[36]

Figure 46: Annual Net Foreign Direct Investment Inflow in Power Sector

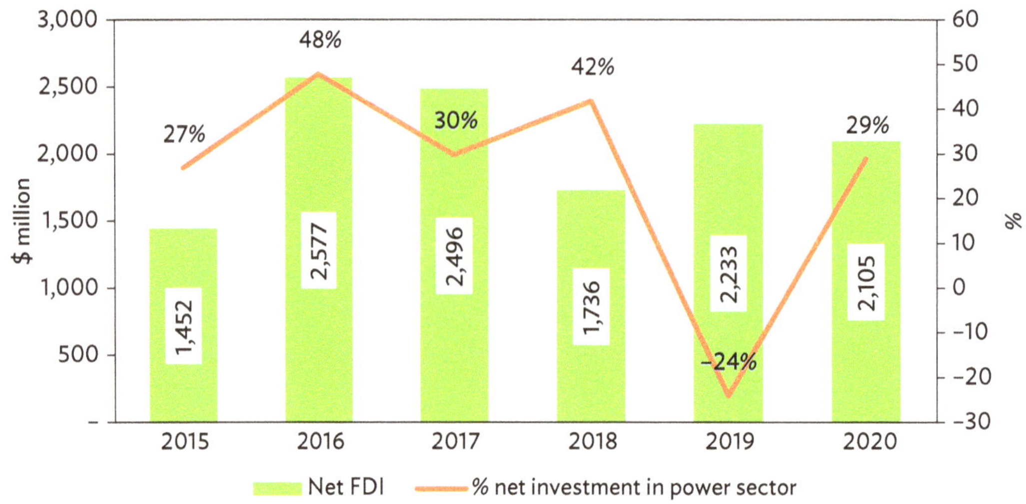

FDI = foreign direct investment.

Source: State Bank of Pakistan.

Pakistan has developed relatively close ties with the People's Republic of China, which has supported the power sector. Some of the power projects, which have received investments as part of China–Pakistan Economic Corridor, are shown in Table 28.

[34] A. Ghouri. 2018. What next for Pakistan's Board of Investment and FDI? LSE Blog. London School of Economics and Political Science. 15 February.

[35] World Bank. 2020. *Doing Business 2020.*

[36] State Bank of Pakistan. 2020. *Annual Report 2019-20*

Table 28: Foreign Direct Investment in Power Projects as Part of the China–Pakistan Economic Corridor

Project Name	Project Details
Sachal 50-megawatt (MW) Wind Farm Project[a]	**Location:** Jhimpir, Thatta District, Sindh Province **Operator:** Sachal Energy Development (private) Ltd. **Total amount invested:** $133.9 million **Total installed capacity:** 49.5 MW **Commercial Operation Date (COD):** 11 April 2017
Sahiwal Coal-Fired Power Plant[b]	**Location:** Sahiwal, Punjab Province **Operator:** Huaneng Shandong Power Generation Co. Ltd. and Shandong Ruyi Science and Technology Group **Total amount invested:** $1.8 billion **Total installed capacity:** 1,320 MW **COD:** 3 July 2017
Suki Kinari Hydropower Station Project[c]	**Location:** Kaghan, Khyber-Pakhtunkhwa **Operator:** SK Hydro **Total project cost:** $1.8 billion **Total installed capacity:** 870 MW **Expected completion date:** December 2022

[a] Belt and Road Initiative. 2018. BRI Factsheet Series – Sachal Wind Farm. 26 October.
[b] Belt and Road Initiative. 2018. BRI Factsheet Series – Sahiwal Coal Power Project. 19 October.
[c] Belt and Road Initiative. 2018. BRI Factsheet Series – Suki Kinari Hydropower Project. 11 October.
Source: Compiled by authors.

These are in addition to the coal-based projects discussed earlier.

Bank Credit and Bonds

Figure 47: Credit to Private Sector for Electricity, Gas, Steam, and Air-Conditioning Supply ($ billion)

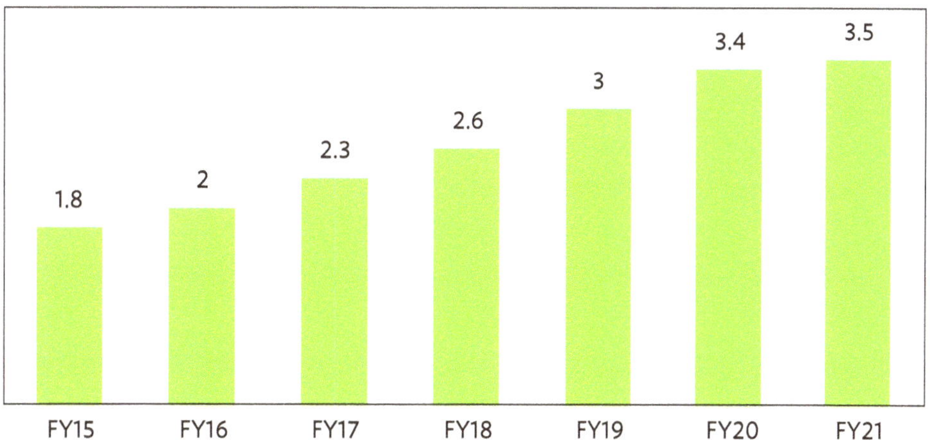

FY = fiscal year.
Sources: Compiled from Pakistan Economic Surveys.

The private energy sector did not have bank credit until 2017 when the State Bank of Pakistan declared an initiative for green credit. However, by March 2021 credit to electricity, gas steam, and air conditioning supply reached 20% of total credit to the private sector (Figure 47).

State Bank of Pakistan formulated a framework for green banking in 2017, defined as a mechanism to help banks and their customers to reduce their carbon footprint and make a transition to a green, low-carbon, and sustainable economy. Although green financing is still at a nascent stage,[37] some of the policy initiatives for green financing are given in Table 29.

Table 29: Financing Renewable Energy in Pakistan

Policy	Description
Scheme for Financing Power Plants Using Renewable Energy (RE) (State Bank of Pakistan [SBP], 2016)[a]	• The first scheme was introduced in 2009 to promote RE projects in Pakistan. However, due to low fund utilization, the state bank revised the framework in 2016. • The new version of the scheme offers debt financing for RE projects at fixed subsidized annual interest rate of 6%. • It is an on-lending structure where the bank lends to commercial banks and other financial institutions at an interest rate of 2%, who further on-lends to end consumers at 6%. The intermediaries bear the loss of defaults.
Green Banking Guidelines (SBP, 2017) - 4.1 3 p15[b]	• The guidelines stress the need for financial institutions to include an evaluation of environmental costs in their credit appraisals to stress the benefits of green alternatives and green finance. • They encourage banks to monitor environmental risks and to include these risk covenants in legal agreements with borrowers.
Advisory agreement between the International Finance Corporation and SBP for green banking (2018)[c]	• Support is provided for the improvement of environmental risk management practices and ethical lending practices.

[a] S. Malik, M. Qasim, and H. Saeed. 2018 Green Finance in Pakistan: Barriers and Solutions. *ADBI Working Paper Series.* No. 880. Asian Development Bank Institute.

[b] State Bank of Pakistan. 2017. *Green Banking Guidelines*.

[c] M. Z. Mumtaz and Z.A. Smith. 2019. Green Finance for Sustainable Development in Pakistan. October; *Dawn*. 2018. WB to support green banking in Pakistan. 31 August.

Source: Compiled by authors.

Some of the banks providing loans in green energy are given in Table 30.

[37] H. Park and J.D. Kim. 2020. Transition towards green banking: role of financial regulators and financial institutions. *Asian Journal of Sustainability and Social Responsibility.* 5(5).

Table 30: Examples of Renewable Energy Financing Schemes

<table>
<tr><th>Financing Scheme</th><th colspan="3">Details</th></tr>
<tr><td rowspan="3">Habib Bank Limited (HBL)[a]</td><td colspan="3">Financing of renewable energy (RE) projects is divided into three categories.</td></tr>
<tr><td>Category I</td><td>Category II</td><td>Category III</td></tr>
<tr><td>• Capacity ranging from more than 1 megawatt (MW) to 50 MW
• Loan amount of up to about PRs6 billion ($0.034 billion) for single project
• Maximum term of 12 years
• Interest rate of 6% per annum or any latest directive from the State Bank of Pakistan (SBP)</td><td>• Capacity up to 1 MW
• Maximum loan amount of PRs400 million ($2.29 million) for single borrower
• Maximum term of 10 years
• Interest rate set at 6% per annum or any new directive from SBP</td><td>• Loan to a certified vendor and supplier under Alternative Energy Development Board Certification Regulations 2018 for installation of wind and solar systems or sale of electricity to owners on a lease basis
• Maximum loan amount of about PRs1 billion ($0.006 billion) for single vendor or supplier
• Maximum term of 10 years
• Interest rate 6% per annum or any new rate set by SBP</td></tr>
<tr><td>Bank Alfala[b]</td><td colspan="3">• Solar energy financing for capacity ranging from 4 kilowatts (kW) to 10,00 kW (about 1 MW) with net-metering
• Loan up to PRs400 million ($2.29 million) to single borrower
• Interest rate of 6% per annum</td></tr>
</table>

[a] Solar On-Grid Solutions. HBL Financing Scheme for Renewable Energy – SBP Approved.

[b] Bank Alfalah. Alfalah Green Energy.

Source: Compiled by authors.

Green Bonds

The bond market in Pakistan was initiated in the 1990s after financial liberalization. However, since then it has developed at a very slow pace. The government has issued securities and bonds such as Treasury bills for short-term borrowing of 3, 6, and 12 months and Pakistan Investment Bonds for tenures of 3, 5, 10, and 20 years.[38]

Green bonds are aimed specifically at financing or refinancing projects that positively contribute to the environment. In June 2021, the Securities and Exchange Commission of Pakistan issued guidelines per section 172 of the Securities Act 2015, which govern the issuance of green bonds. Green Bond Principles require issuers to promote transparency, accuracy, and disclosure of green bonds to help investors understand their characteristics.[39] Green bonds in Pakistan are required to be issued at a minimum value of PRs5,000 ($32.27) per certificate or multiples thereof.[40] The projects eligible for green bond investment are as follows:

[38] S. Ahmed. 2021. Bond market primer. *Dawn*. 23 March.

[39] ICMA. 2018. *Green Bond Principles*. Paris.

[40] Securities and Exchange Commission of Pakistan. Guidelines - Issuance of Green Bonds/Sukuks.

- Renewable and sustainable energy
- Energy efficiency
- Pollution prevention and control
- Environmentally sustainable management of living natural resources and land use
- Terrestrial and aquatic biodiversity conservation
- Clean or environment-friendly transportation
- Sustainable water management
- Climate change adaptation
- Green buildings
- Eco-efficient, circular economy-adapted products and or any other green project

Table 31 gives examples of recent green bonds issued by the government. It is hoped that the issuance of these bonds will raise environmental awareness among financial lending institutions and encourage them to enforce standards on environmental, social, and governance issues across their portfolio.[41]

Table 31: Examples of Green Bonds

Issuer	Description
Pakistan Water and Power Development Authority (WAPDA) with JP Morgan Chase & Co. advising[a]	**Amount:** $500 million green bond **Issue date:** 2021 **Coupon rate:** 7.5% **Term:** 10 years **Use of proceeds:** Renewable energy and climate change adaptation projects **Listing:** London Stock Exchange **Description:** This is a debut green bond issued for the development of hydroelectric power projects, namely the Diamer Basha and Mohmand dams. WAPDA green bonds were oversubscribed by 6 times, reflecting global interest in green project investments.
Power Holding Private Limited- Power division of Ministry of Energy[b]	**Amount:** PRs200 billion ($1.3 billion) Pakistan Energy Sukuk **Issue date:** 2019 **Term:** 10 years with semiannual payments **Listing:** Pakistan Stock Exchange **Description:** The bond is issued to help the government resolve the debt crisis and ease the liquidity scarcity in the energy sector (that includes oil and gas suppliers), distribution companies, the Water and Power Development Authority, and power producers.

[a] *The Business Standard.* 2021. Pakistan plans to raise $500 million from first green bond sale. 22 May; T. Gillespie and G. Ritchie. G. 2021. Pakistan Plans First Green Bond to Fund Hydropower Projects. *Bloomberg.* 18 February; and JP Morgan Chase & Co. 2021. *Development Finance Instution Annual Report 2021.*

[b] *Dawn.* 2019. Ministry of Energy issues Rs200bn 'Pakistan Energy Sukuk' 2 March.

Source: Compiled by authors.

Other smaller sources of funding for clean energy initiatives include corporate social responsibility allocations of private sector firms, Clean Development Mechanism allocations and microfinance loans.

[41] Y. Herekar. 2020. Pakistan and Green Bonds: A Case Study. LinkedIn Blog. 14 September.

International Financial Assistance: Multilaterals and Bilaterals

Multilateral development banks are important for sustainable finance initiatives in the country. Their total climate finance to Pakistan between 2015–2019 was approximately $5.4 billion, of which the total funding to the energy sector was estimated to be around $1.9 billion (Figure 48).[42]

Figure 48: Year-Wise Investment in Climate Finance for Energy Sector ($ million)

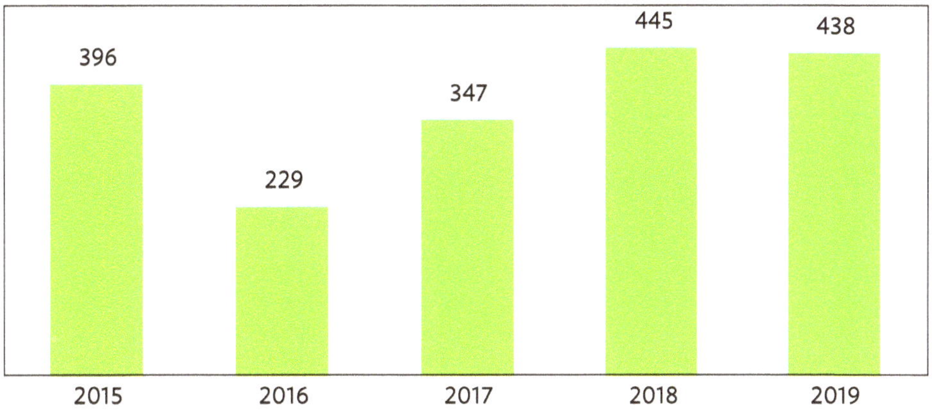

Note: Multilateral development banks cover combined data from the Asian Development Bank, the European Bank for Reconstruction and Development, the European Investment Bank, the Inter-American Development Bank Group, the World Bank Group, and the Islamic Development Bank. It does not include funding from KfW, Deutsche Gesellschaft für Internationale Zusammenarbeit, and Department for International Development, United Kingdom.

Source: 2019 Joint Report on Multilateral Development Banks Climate Finance.

Some examples of the support provided by multilateral agencies for the development of renewable energy in the country are in Table 32.

Table 32: Examples of Multilateral Agencies support for Renewable Energy Projects

Investor	Details
International Finance Corporation (IFC)	
Development of incremental renewable energy generation projects for approximately 12 million people.[a]	• Amount: $755 million • Installed capacity: 2 hydro projects (820 megawatts [MW]), 4 wind projects (300 MW) and the country's first liquefied natural gas terminal.
Karot hydropower project	• Amount: $100 million • Installed capacity: 720 MW • Date of commissioning: 2017 • Investment made in Karot hydropower project[b] using the investment platform China Three Gorges South Asia. The location of the project is the river Jhelum, about 55 kilometers (km) east of Islamabad.[c]

continued on next page

[42] The value is the sum of MDB adaptation and mitigation finance, estimated as a share of the total for the South Asia region using the same proportion as Pakistan's share in the total climate finance of South Asian countries.

Table 32 *continued*

Investor	Details
World Bank Group	
Sindh Solar Energy Project	• Amount: $100 million • Installed capacity: 2,160 MW • Date of commissioning: 2017 • Sindh Solar Energy Project to support further private investment in utility-scale and off-grid solar and the second stage of the Dasu Hydropower project.[d]
The Khyber-Pakhtunkhwa Hydropower (K-P) and Renewable Energy Development Project (2020)[e]	• Amount: $450 million (mix of commercial and concessional financing)–$250 million of International Development Association (IDA) and $200 million of International Bank for Reconstruction and Development loan (IBRD). • Details: Investment in hydroelectric power and solar in K-P province to strengthen institutions in the energy sector to support the transition to clean energy.
Pakistan Programme for Affordable and Clean Energy (PACE) (2021)	• Amount: a) $600 million–$300 million from the IDA and remaining $300 million as IBRD loan. b) Approved $400 million (footnote a). • Details: Reduction in circular debt through cost cutting in power generation, decarbonizing the energy mix, improving distribution by increasing the participation of the private sector and retargeting electricity subsidies and tariffs for consumers.[f]
Green Development Program Project[g]	*Financing detail* • Project description: This project is meant to improve environmental governance and promote green investments in the Punjab. • Total cost: $273 million • Commitment amount: $200 million *Timeline* • Approved in fiscal year (FY) 2018 • Closing date 30 May 2025 *Sectors* • Other public administration: 63% • Public administration (Energy and Extractives): 10% • Other industry, trade, and services: 14% • Other transportation: 8% • Waste management: 5%
Tarbela Fourth Extension Hydropower Project[h]	• Project description: This included building a powerhouse, modifying a tunnel, and installing three 470 MW power plants of 1,410 MW in total on an existing dam located 60 km northwest of Islamabad on the main River Indus. • Implementing Agency: Water and Power Development Authority (WAPDA) and National Transmission and Despatch Company (NTDC) • Date: The project was approved in 2012 and was expected to be completed in 2022. • Total cost: $914 million • Commitment amount: $840 million
Dasu Hydropower Stage I Project[i]	• Project description: Setting up of a 2,160 MW hydroelectric plant on the river Indus, which can be increased to 4,320 MW in the future. • Implementing Agency: WAPDA and NTDC • Date: The project was approved in 2014 and is expected to be completed in 2024. • Total cost: $3,787.70 million • Commitment amount: $588.40 million

continued on next page

Table 32 *continued*

Investor	Details
Sindh Solar Energy Project[j]	• Project description: The project aims to increase solar power generation and access to electricity in Karachi, Hyderabad, and other districts of Sindh Province. • Implementing Agency: Energy Department, Government of Sindh • Date: The project was approved in 2018 and is expected to be completed in 2023. • Total cost: $105 million • IDA Commitment amount: $100 million

[a] World Bank. 2019. Pakistan: Powering Homes and Businesses. *Brief.* 2 January.

[b] It was Pakistan's largest private renewable energy project, and IFC's largest hydroelectric power project as of 2018.

[c] World Bank. Karot Hydropower Plant.

[d] The total cost of the project is estimated to be $782 million, of which $450 million is provided by the World Bank; $223 million is to be borrowed from local commercial banks, export credit agencies and private sources; and the remaining $134 million is to be arranged by the government.

[e] *The Express Tribune.* 2020. World Bank approves $450m loan 26 September; World Bank. 2021. World Bank Board Approves New Financing Totaling $800 Million to Support Critical Power Sector Reforms and Human Capital Development. Press Release. 28 June.

[f] M. Haider. 2021. Pakistan seeks $1.2 billion World Bank program loan for power sector reforms *The International News.* 1 May.

[g] World Bank. Punjab Green Development Program.

[h] World Bank. Tarbela Fourth Extension Hydropower Project.

[i] World Bank. Dasu Hydropower Stage I Project.

[j] Powerline. 2019. World Bank and Pakistani government sign $100 million financing agreement.

Source: Compiled by authors.

In terms of major donors, the government has taken a loan of $200 million from the World Bank under the Green Development Program to strengthen environmental governance and promote green investment in Punjab.[43] Currently, the World Bank has 58 active projects worth $14.8 billion of which 24% are in the energy and extraction sector.[44] The Asian Development Bank (ADB) has also supported the energy sector. As of 18 March 2021, ADB had funded projects totaling approximately $9.53 billion in the energy sector, which is 26% of total value of projects and technical assistance to Pakistan from ADB.[45] In May 2021, the government and ADB finalized a $300 million loan to finance a 300 MW hydroelectric plant in Balakot, Khyber-Pakhtunkhwa province.[46]

Of the national agencies the Japan International Cooperation Agency (JICA)[47] and the United States Agency for International Development are the major bilaterals supporting Pakistan.[48] For example, in 2016 and 2017, JICA invested ¥5,000m[49] ($46 million) and ¥2.665 billion ($23.76 million)[50] for the Energy Sector Reform Program (II) and the Islamabad and Burhan Transmission Line Reinforcement Project (Phase 1), respectively. United States Agency for International Development has been active

43 World Bank. Pakistan-Punjab Green Development Program.

44 World Bank. Maps.

45 ADB. N.D. Pakistan: Cumulative Commitments.

46 Energy Central. 2021. Banking: ADB Approves $300 Million Loan for Hydropower Plant to Boost Clean Energy in Pakistan. 30 March.

47 As per official development assistance (ODA) loan project data, the latest JICA ODA loans project toward Pakistan is seen until 2017 (meaning of this footnote is unclear and can be left out).

48 JICA. ODA Loan Project DATA.

49 ¥1 = $0.0092 average 2016 rate. Source: ExchangeRates. Japanese Yen to US Dollar Spot Exchange Rates for 2016.

50 ¥1 = $0.0089 average 2017 rate. Source: ExchangeRates. Japanese Yen to US Dollar Spot Exchange Rates for 2017.

in support of renewable energy since 2010. This has included technical assistance loan guarantees to encourage private investment, advisory services to promote smart technology and the construction of transmission lines and substations to evacuate wind-generated power to the national grid, as well as the rehabilitation of thermal and hydroelectric power plants.[51]

Support for Energy Efficiency

The National Energy Efficiency and Conservation (NEECA) Act 2016 currently provides the governance framework to promote energy-efficient practices in all sectors. NEECA was created under this act to plan for future energy-saving initiatives.[52] NEECA has set out a road map to save up to 3 Mtoe by 2025 (base year 2020) by implementing energy efficiency and conservation programs in all major sectors of the economy.[53]

The institutionalization phase of the roadmap began in FY 2020, with the formation of the National Energy Efficiency and Conservation Policy and the Energy Efficiency and Conservation Tribunal, the establishment of designated provincial agencies, and the creation of provincial action plans. The operationalization phase began in FY 2021, with the Standard and Labelling and Minimum Energy Performance Standards system for monitoring, verification, and enforcement, and the accreditation of auditors and laboratories. The system aims to give a level playing field to manufacturers, as well as a quality product to customers. It also includes a complaint response mechanism.

Examples of NEECA initiatives for demand-side energy efficiency are standards and labeling arrangements for efficient fans, saving approximately 25 MW of electricity; Zig-Zag technology for brick kilns, which has the potential for 30%–40% savings in energy usage in the brick kiln industry;[54] and a 5-star energy labeling scheme for energy-efficient LED lighting, which has the potential to save 1 million tons of CO_2 emissions, 1.2 TWh of electricity, and more than $120 million in electricity costs by 2030.[55]

The implementation phase is expected to draw on various sources of funding ranging from the Public Sector Development Program, donor assistance, a revolving loan fund, commercial banks, microfinance institutions, and from an energy efficiency surcharge (Table 33).

51 United States Agency for International Development. Energy.

52 NEECA. N.D. Overview: About NEECA.

53 NEECA. 2020. Strategic Plan 2020-2023.

54 Government of Pakistan, Planning Commission. 2018. National Action Plan Sustainable Energy for All. United Nations Development Programme Pakistan.

55 United for Efficiency. 2021. Successful Implementation of First Minimum Energy Performance Standards (MEPs) and Labels for Energy Efficient LED Lighting in Pakistan. 8 January.

Table 33: Sources of Funding for Energy Efficiency Implementation

Sector	Description/ Major Activities	Funds Requirement	Performance Targets
Industry (1.3 Mtoe)	• Implementation of Minimum Energy Performance Standards (MEPS), mandatory energy audits through certified energy auditors/managers, energy-saving certificate schemes, and energy efficiency financing mechanisms.	PRs370 million ($2.3 million)[a] Potential Sources: • Public Sector Development Programme (PSDP) • Revolving loan fund (RLF) • Donor assistance • State Bank of Pakistan – Green Banking Guidelines • Microfinancing	• Energy audit reports for 200 textile industrial units will be available by 2023. • Energy savings of up to 0.1 million tons of oil equivalent (Mtoe) in sugar mills by 2023. • Variable - frequency drive on pumps and motors installed in 500 industrial units by 2023 to save 0.3 Mtoe.
Building (0.5 Mtoe)	• Implementation of building codes and mandatory labeling schemes in all new construction in domestic, commercial, and public sectors.	PRs310 million ($1.9 million)[a] Potential Sources: - • RLF • Donor Assistance • PSDP • Commercial Banks • Microfinancing	• Naya Pakistan Housing Authority will implement Energy Efficiency Building Codes. • Coordination mechanism established with the Ministry of Climate Change and provincial departments for distribution of energy-efficient cooking stoves will help save 0.08 Mtoe by 2023. • Development of NEECA registry portal for energy-efficient fans.
Transport (0.5 Mtoe)	• Regulating vehicle tuning centers, the vehicles examination system, fleet management mechanisms and supporting the implementation of the Electric Vehicle Policy 2019.	PRs120 million ($0.7 million)[a] Potential Sources: - • PSDP • Proposed energy efficiency surcharge • Microfinance • RLF • Donor assistance	• Coordination established between National Highways Authority and Provincial Transport Authorities. • Energy performance standards established for vehicle emissions.
Power (0.4 Mtoe)	• Introduction of mandatory energy standards and labeling schemes.	PRs1,800 million ($11 million)[a] Potential Sources: - • PSDP • Energy efficiency surcharge (proposed) • Donor Assistance	• Establishment of coordination mechanism with DISCOs and K-Electric for installation of smart meters. • Installation and upgrading of electricity grids will save up to 0.1 Mtoe by 2023. • Replacement of copper wire transformers with smart/ efficient transformers, identification of overloaded feeder and program development for power factor improvement.

continued on next page

Table 33 *continued*

Sector	Description/ Major Activities	Funds Requirement	Performance Targets
Agriculture (0.3 Mtoe)	• Replacing inefficient tube wells, establishing tractor tuning centers, minimum performance standards for motors, and training programs for stakeholders	PRs600 million ($3.7 million)[a] Potential Sources: - • PSDP • Energy efficiency surcharge (proposed) • Microfinance • RLF • Donor assistance	• Launch of MEPS for motors and mandatory tubewell labeling. • Development of National Energy Efficiency and Conservation Authority Registry Portal for energy-efficient electric tube wells. • Establishment of automotive engine diagnostic and tuning centers.

[a] $1 = PRs 162 (Jan 2011–Nov 2021 average).
Source: NEECA 2020.

Energy service companies (ESCOs) are needed to implement country-level energy services like installation and supply of energy-efficient equipment, retrofitting and process improvement. ESCOs are still in their infancy in the country and there are only a few domestic companies working to finance energy efficiency measures and to provide energy services (footnote 53).

There have been some limited initiatives to support these. For example, NEECA has supported energy efficiency improvements of water pumping systems.[56] In 2016, the Small and Medium Enterprises Development Authority in partnership with JICA developed an ESCO manual that highlights the process of energy audit and energy efficiency.[57] Small and Medium Enterprises Development Authority has collaborated with international partners such as the German International Cooperation agency and Training and Development Centres of the Bavarian Employers' association to provide services like energy audits and training programs.[58] The Energy Conservation Fund, in collaboration with banks and financial institutions, offers concessional loans to small and medium-sized enterprises for the promotion of energy-efficient technologies.[59]

Examples of some of the energy efficiency projects in the country are given in Table 34. However, much remains to be done if the ambitious projections for demand-side savings in the low-carbon scenario are to be achieved.

[56] NEECA. N.D. The role of NEECA.
[57] Japan International Cooperation Agency. 2013. ESCO Manual.
[58] Small and Medium Enterprise Development Authority. n.d. Energy Efficiency Initiatives. *Business Facilitation*.
[59] NEECA. 2022. INNOVATIVE FINANCING FOR ENERGY EFFICIENCY.

Table 34: Examples of Energy Efficiency Projects

Projects	Description
Sitara Chemical Cooling Tower Project[a]	• **Project:** Sitara Chemicals Industries Limited uses 4 pumps of 93-kilowatt (kW) capacity each for the cooling tower. The company approached Enersave, an energy service company (ESCO), to conduct a study to improve the operation of the cooling tower. • **Achievement:** After a detailed study, the company implemented an effective cooling tower management system, which resulted in reduction of 1 pump of 93 kW capacity saving 1 million units of electricity per year.
Renewable Energy and Energy Efficiency Project[b]	• **Description:** Project aims to increase the renewable energy (RE) in the total energy mix. • **Time period:** 2005 to 2023 (second phase started in 2014) • Implementing Agency: German Federal Ministry of Economic Cooperation and Development (BMZ) • **Target area:** Grid management, distributed generation, rural electrification, and Pakistan German Renewable Energy Forum. • **Some of the Achievements:** – The project provided policy advisory support to the Alternative Energy Development Board (AEDB), resulting in the addition of over 2 gigawatts (GW) of RE to the national grid. – Over 1,500 quality-assured solar home systems have been installed in off-grid areas, benefiting an estimated 10,000 people. – Supported State Bank RE refinancing schemes. As of December 2019, PRs8.5 billion ($56.4 million)[c] had been released for 93.6-megawatt (MW) capacity across 13 projects. – It also supported the State Bank of Pakistan in implementing in the Green Banking Guidelines.
EE Lighting project[d]	• **Project detail:** It focuses on establishing efficient lighting policies and creating a framework for financial mechanisms that support innovative and high-efficiency products for climate change mitigation and securing environmental benefits. • **Time period:** 2017 to 2021 • **Implementing Agency:** UN Environment Programme (UNEP) • **Executing Agency:** NEECA, Ministry of Energy (Government of Pakistan) and International Institute for Energy Conservation (IIEC) • **The United Nations Development Assistance Framework Linkages:** To strengthen the capacity of public and private sector entities to implement policies and plans in energy efficiency and RE. • **IIEC function:** To provide expertise and technical assistance to support the development of the National Efficient Lighting Strategy, design a revolving loan fund for financing energy-efficient lighting products, and implement pilot demonstration projects on LED lighting. • **Total funding of $7.38 million (Global Environment Facility [GEF] grant and cofinancing)** – GEF project grant: $1.58 million – Cofinancing total: $5.8 million – Total disbursement (June 2020): $1.3 million – Total expenditure (June 2020): $413,22

[a] Enersave. Conserving Energy, Minimizing costs.
[b] GIZ. Promoting renewable energy and energy efficiency in Pakistan.
[c] $1 = PRs151 (2019 average).
[d] UNEP GEF PIR Fiscal Year 2020; IIEC. Energy Efficient Lighting Market Transformation in Pakistan.
Source: Compiled by authors.

In summary, the low-carbon scenario requires large amounts of investment and technical support if its figures are to be achieved. This investment and support will not be forthcoming without significant effort from both the government and the private sector. If addressing climate change is to be given the priority implied by recent policy announcements it will require a major drive to both facilitate funding and to align incentives for the private sector with this priority.

CHAPTER 5

CONCLUSIONS

As Pakistan is committed to revive its economy from the effects of the pandemic, natural calamities, and global events, it aspires to become an upper middle-income economy by 2047 when it completes 100 years of its independence. This socioeconomic transformation will exert high energy demand from various economic sectors. Moreover, the country is highly vulnerable to the adverse effects of climate change, and it is among the top 10 countries around the world to be affected by climate change. In 2016, under its NDCs to address climate change, the government proposed to reduce GHG emissions by up to 20% in 2030 taking the base year as 2015. This commitment was updated in 2021, where the government targeted an overall reduction of 50%, of which 15% is unconditional and remaining 35% is conditional on the receipt of adequate international assistance.

To assist this process, this study develops a technological road map for facilitating a low-carbon transition in Pakistan. This road map discusses energy options and the current barriers and enablers to their development. Critically, there are large funding requirements for a low-carbon growth scenario and the report gives estimates of the investment required and discusses potential funding sources. The additional funding required is substantial and success will necessitate active government involvement, as well as support from the private sector.

The technological road map has been developed using a robust modeling-based approach. The model used is the MESSAGEix model developed by the International Institute for Applied Systems Analysis (IIASA). A detailed bottom-up scenario-building exercise has been performed to develop two alternative scenarios, one based on current trends (BAU scenario) and the other a low-carbon trajectory, required to meet the government's climate change ambitions (low-carbon scenario). The low-carbon scenario examines a low-carbon energy and emissions trajectory based on policies directed toward achieving the Paris Agreement goals and Pakistan's own agreed commitments. This scenario, by definition, is not purely a cost-optimization exercise, as a level of commitment to emissions reduction is built into the analysis. In the process of obtaining the least-cost low-carbon pathway and improving the robustness of the scenario, sensitivity analysis has been conducted for several important parameters such as cost, efficiency, and different technological capacity targets.

Both BAU and low-carbon scenarios use the same assumptions regarding income and population growth, but the BAU scenario follows past and current trends in low-carbon policy and technological interventions, while the low-carbon scenario assumes significant and rapid deployment of emerging low-carbon technologies and options to reduce GHG emissions. A comparison of the investment requirements of the two scenarios shows that the low-carbon scenario would require an additional investment of $191.8 billion (in 2022 prices) between 2020 and 2050 over the BAU scenario, so the investment commitment is substantial. Energy sector GHG emissions rise in both scenarios, but under the low-carbon scenario, total energy sector GHG emissions would be almost 23% lower in 2030 and 36% lower in 2050, as compared with those under the BAU scenario.

The three key pillars of the decarbonization strategy are improvements in energy efficiency, fuel switching to low-carbon fuels, and electrification.

Energy efficiency improvement has a critical role in reducing demand. With the increasing adoption of energy-efficient technologies, processes, and appliances, final energy requirements from user sectors are expected to reduce over time, resulting in energy and emissions savings. Additionally, from the supply side, a switch toward more efficient power production will also contribute to the reduction in energy demand.

TFED is estimated to be 12% lower under the low-carbon scenario by 2050. Out of the total energy demand reduction of about 51 Mtoe in 2050, the main reductions are in industry (47% of the total) and transport (39%). The low-carbon scenario shows a strong effect from the application of energy efficiency measures and the majority of the emissions mitigation in the low-carbon scenario relative to that in the BAU scenario (91% in 2030 and 75% in 2050) comes from demand-side initiatives by users. In the low-carbon scenario, the key mitigation opportunities from the demand-side lie in the coal-to-gas transition in industry, electrification of transport, and gas and electric- based cooking. From the supply side the key areas with high mitigation potential are nuclear-based power generation, hydropower generation, and solar PV and wind power generation.

Figure 49: Primary Energy Supply Under the Business-as-Usual and Low-Carbon Scenarios

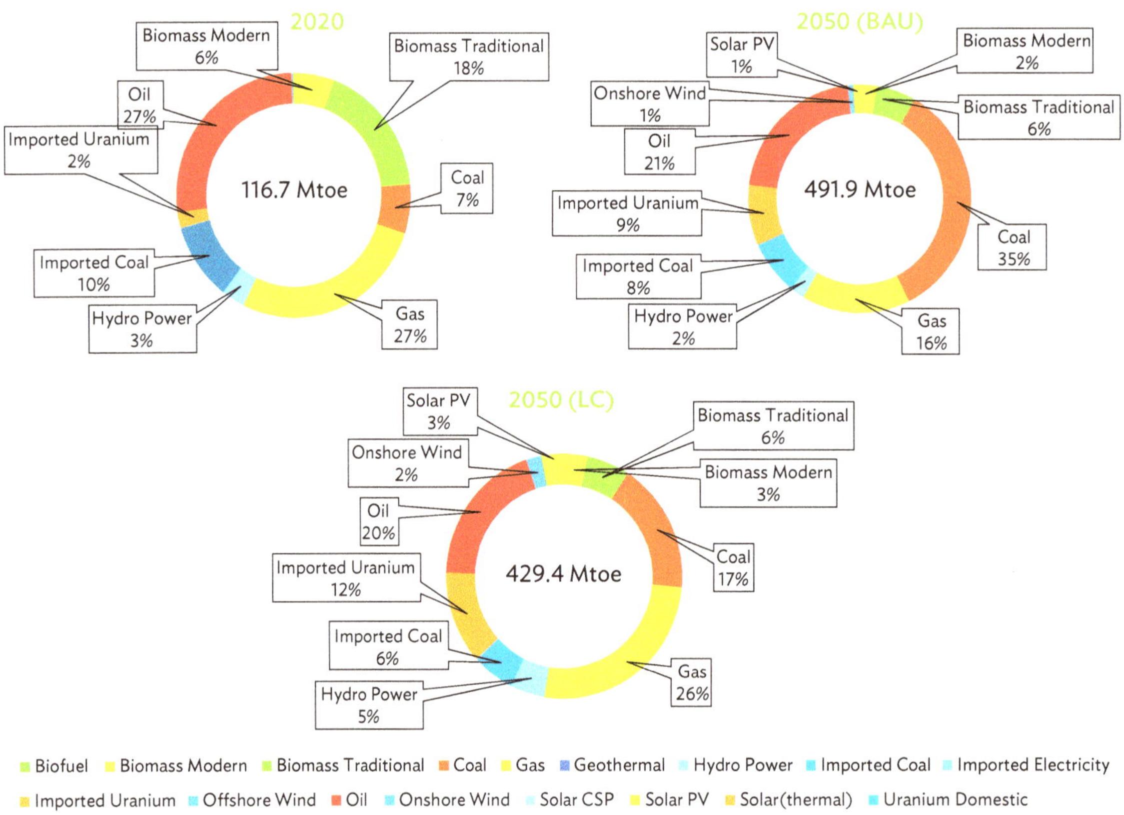

BAU = business-as-usual, CSP = concentrated solar power, LC = low carbon, Mtoe = million tons of oil equivalent, PV = photovoltaic.
Source: Authors.

Fuel switching toward cleaner alternatives, like nuclear, renewables like solar and wind, and natural gas is also an important part of the strategy. The role of nuclear power, renewables, and natural gas is crucial to support the clean energy transition. Under the low-carbon scenario, around 26% of TPES would come from natural gas and about 22% collectively from nuclear fuels, solar, hydro, and wind by 2050. The share of imported uranium in primary energy would increase from 7% in 2030 to 12% in 2050. Natural gas would be the dominant fuel in the PES mix between 2030 and 2050 with a share of 27% in 2030 to 26% in 2050 (Figure 49). It would be the major fuel imported in the low-carbon scenario, followed by oil and uranium. This is chiefly because the energy transition in a range of sectors from coal, biomass, and oil to natural gas, would lead to an accelerated gas demand and domestic reserves are fully depleted by the end of the period. The share of imports in PES is higher in the low-carbon scenario and this would require careful planning to minimize geopolitical risks to the country's energy security.

In terms of emissions mitigation, in the low-carbon scenario, the key areas with high mitigation potential are nuclear-based power generation, hydropower, and solar PV and wind power. Nuclear energy would play a crucial role in meeting growing electricity demand with negligible emissions. Generation from nuclear fuel is a mature technology, however, the safety aspects including proper transport and disposal of nuclear fuel require careful planning. In the low-carbon scenario, the installed capacity of nuclear power plants is estimated to increase from 3.1 GW in 2025 to 36 GW in 2050. However, although, the country has ambitious targets for capacity additions by nuclear power plants, the high investment costs, and safety issues with nuclear fuel are obvious barriers that need to be overcome if these targets are to be achieved.

Despite being only recently introduced in the energy sector, coal has a significant share in the generation mix. Almost 19% of the power generation in the country came from just four coal-fired China–Pakistan Economic Corridor power plants 2019–2020 (footnote 21). The phasing out of this dependence on coal in the power sector is a key part of the road map. In the low-carbon scenario, installed capacity of coal powered plants decreases from 9.3 GW in 2030 to 7.8 GW in 2050 and coal consumption grows by 9% annually between 2030 and 2050 primarily owing to its role in industry sector and preexisting power plants. The key enabler for coal-based power generation in the country is its low cost, however a coal-based power generation system has a high emissions content and lacks the flexibility to support a high level of renewable energy in the grid, which must be provided by other technology options. Currently natural gas-based power generation costs are higher than for coal-based power, however natural gas power plants offer the opportunity to balance the variability of increasing renewable power generation within the grid. In addition, emissions per kWh of electricity generation from gas-fired power plants is estimated to be 60%–70% lower than in comparable coal power plants. In the low-carbon scenario natural gas acts as a key supporting fuel in the transition from coal-based power generation.

Electrification is also important for the energy transition given the potential for efficiency gains and decarbonization in user sectors. However, for electrification to be a successful strategy in meeting the goal of decarbonization, it is essential to move away from fossil-based power generation to a system where renewables are much more important. A rapid expansion in non-fossil-fuel electricity requires major investment in nuclear and renewable sources, hydro, solar, and wind. Under the BAU scenario, the country continues to invest in capacity based on natural gas and coal (with 26% of total capacity in 2050) with moderate capacity additions in renewable energy-based generation. Under the low-carbon scenario, there is a major restructuring in the capacity mix, with a significant push toward solar PV, onshore wind, and hydro-based power generation. Together under the low-carbon scenario these take 61% of power generation in 2050 and 81% of installed capacity (Figures 50 and 51).

Figure 50: Electricity Generation in Terawatt-Hours Under the Business-as-Usual and Low-Carbon Scenarios

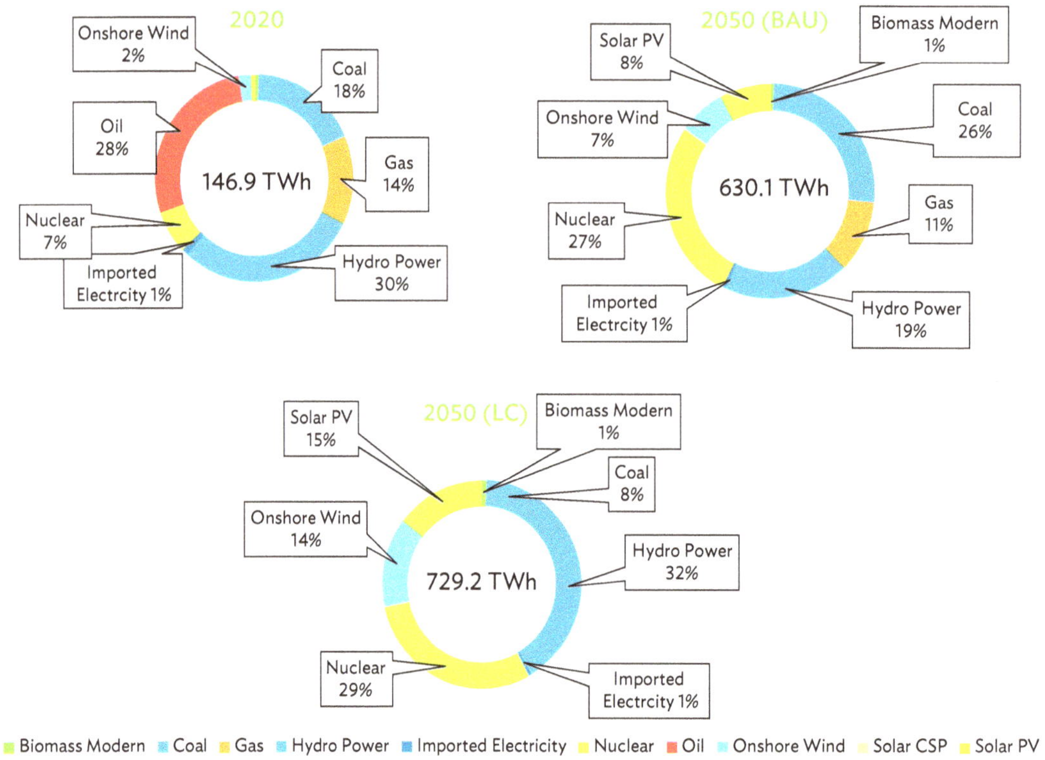

BAU = business-as-usual, CSP = concentrated solar power, LC = low carbon, PV = photovoltaic, TWh = terawatt-hour.

Source: Authors.

Figure 51: Power Generation Capacity in Gigawatts Under the Business-as-Usual and Low-Carbon Scenarios

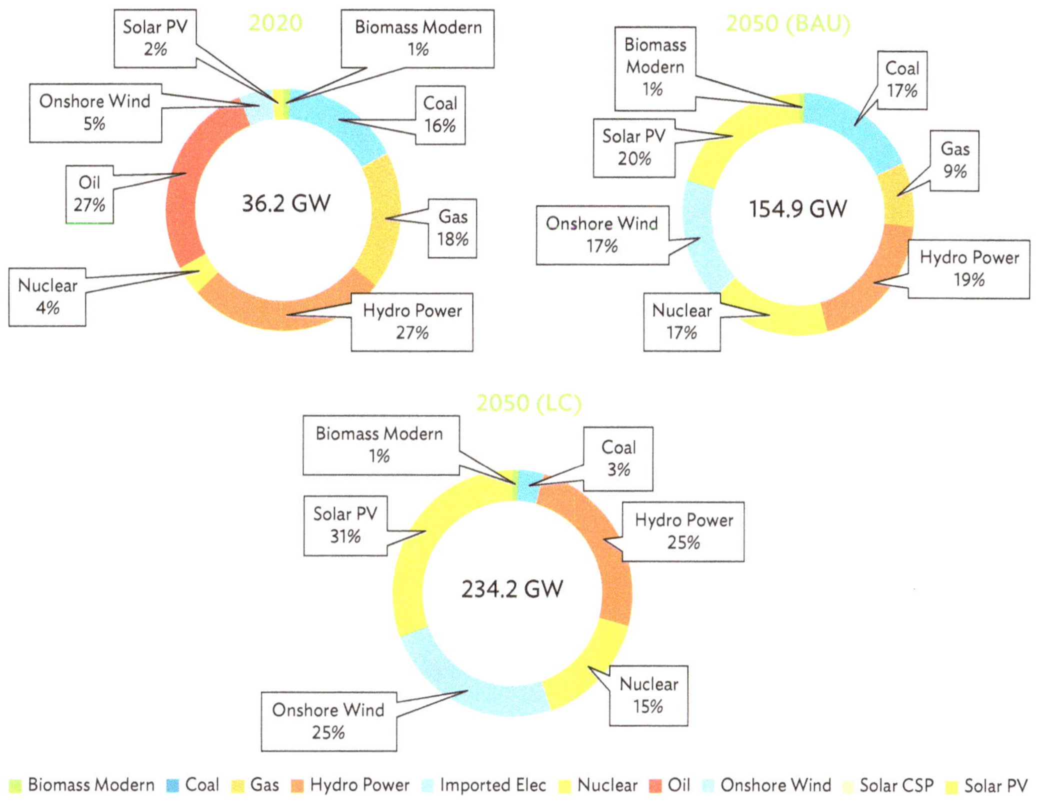

BAU = business-as-usual, CSP = concentrated solar power, GW = gigawatt, LC = low carbon, PV = photovoltaic, TWh = terawatt-hour, PV = photovoltaic.

Source: Authors.

In the low-carbon scenario, hydropower is the largest source of electricity generation by 2050. The potential for hydropower development in Pakistan is vast and has been estimated to be 50–60 GW.[60] Out of this total only 7,320 MW has been developed, so far. While the investment cost of large-scale hydropower plants is high, the long technical and economic life of the system can result in a relatively low unit cost of electricity generation. Hydropower generation can also provide flexibility to the electricity grid where, as in the low-carbon scenario, there is a high share of variable renewable energy. In addition to large hydropower plants, small hydropower generation also provides an opportunity to generate clean power and mitigate emissions. In the low-carbon scenario, hydropower is 25% of installed capacity and provides 32% of electricity generation in 2050. However, there are also constraints on hydro development. Large-scale hydropower generation requires a high investment and a large land area. The reservoir can submerge fertile land and require significant resettlement of

60 International Hydropower Association. South and Central Asia; and IAEA. 2018. Country Nuclear Power Profiles Pakistan.

communities, creating social barriers to the acceptance of hydropower development and possibly an unwillingness by international donors to finance this type of project.

Solar and wind sources also see a substantial expansion of their capacity in the low-carbon scenario. As a mature technology, with increasing capacity additions globally, the development cost of solar PV power plants has come down significantly in the last decade. With reduction in the initial investment cost, the cost of power generation has decreased and is expected to reach a par with coal and hydro-based power generation in the coming years. However, despite the abundant solar energy potential in the country, particularly in Sindh, Balochistan, and some southern parts of Punjab, the intermittency and variability of solar energy is a barrier to further development. Further, current plans and policies, like the National Electric Power Regulatory Authority Act, National Electricity Policy, Alternative and Renewable Energy Policy 2019 and the National Electricity Plan 2023–2027 do not focus on development of energy storage technologies. Therefore, Pakistan needs to make concerted efforts in making provisions for legal, policy, and regulatory support toward development of viable energy storage technologies like pumped hydro and battery storage.

Nonetheless in the low-carbon scenario, solar PV power generation capacity is projected to grow from 6 GW in 2025 to 72 GW by 2050, which is a high annual rate of 11%. Similarly, for onshore wind the low-carbon scenario identifies a significant scope for wind-based power generation, with installed wind power capacity reaching 57.6 GW by 2050, implying an annual growth of 12% between 2025–2030, with a share in generation of 14%.

While the renewable energy-oriented low-carbon scenario is beneficial for the environment and improving energy security, Pakistan could also benefit from local manufacturing of the renewable energy technologies. The Alternative and Renewable Energy Policy 2019 has a local content and indigenization element for promoting the local manufacturing of solar PV system parts or components with multiple synergies like foreign exchange savings, human resource development, and job creation.[61]

The expansion plans as envisaged under low-carbon scenario imply major investment commitments. In low-carbon scenario, hydropower generation requires $153 billion, nuclear power requires $103 billion, wind power requires $62 billion, and solar $51 billion between 2020 and 2050. To handle growing power demand and maintain grid stability, investment in transmission and distribution of $22 billion is also required. These large investments in the power sector are in addition to the investment required in transport and the domestic sector to achieve energy efficiency savings.

Achieving such an ambitious investment program will be challenging. Given the country's current macroeconomic condition, the priority is more on economic growth than moving toward a low-carbon trajectory. Therefore, moving away from coal toward more expensive renewable energy technologies may be challenging for the country. Given this condition, continuity of coal seems an inevitable choice, and it is expected that Pakistan will continue to rely on coal until full economic recovery happens. Direct government funding is likely to be limited, so a majority of funds will have to come from the foreign and domestic private sector—though direct investment, bank credits, and bond issues—and international financial assistance. This will require major policy initiatives to strengthen the system of financial intermediation, to align incentives for the private sector with the goals of environmental policy, ensure the regulatory system supports these goals effectively and develop a project preparation

[61] Climate Change Laws of the World. Alternative and Renewable Energy Policy 2019. Pakistan.

and implementation system that meets the requirements of multilateral and bilateral donor organizations. Only if there is major progress in these areas would it be realistic to think that the low-carbon scenario outlined in the study can be fully funded and implemented. In addition, from the side of energy delivery, effective system planning, adequate mechanisms and incentives for supply- and demand-side flexibility, short-term system balancing, and stability procedures will be required to support a transition to a low-carbon pathway (IEA 2021).

APPENDIX

Decarbonization Beyond the Low-Carbon Scenario

A key aspect of technological development not covered in the low-carbon scenario relates to the commercial development of hydrogen as a green fuel. While globally the current hydrogen demand is met by gray hydrogen, with technology development, the hydrogen market can become green and carbon-free over time. Based on the current state of research on the technology, there is a potential for hydrogen to replace conventional fossil fuels in areas including the following:

- Hydrogen-blended natural gas supply
- Industrial demand for ammonia production, oil refinery, steel production, cement production
- Hydrogen fuel cell (HFC) vehicles
- Hydrogen-based energy storage

To have a fuller decarbonization, options for future emissions mitigation not considered in the model are listed below, for consideration in the future.

(i) **H-CNG.** Hydrogen-blended compressed natural gas (CNG) is the mix of hydrogen and CNG with the share of the hydrogen level going up to 30%.[1]

(ii) **DRI-Hydrogen.** Green hydrogen-based DRI process to make green steel. Green hydrogen-based DRI technology has a TRL 6-8,[2] with an expectation of commercial availability by 2030.

(iii) **Green hydrogen for ammonia production and refinery application.** Replacement of the existing gray hydrogen supply with the green hydrogen to reduce the overall emissions.

(iv) **Hydrogen in cement production.** Blending of hydrogen in the cement production fuel mix. As some process-based emission from cement production is inevitable, emissions from fuel consumption can be minimized using hydrogen as a fuel.

(v) **Hydrogen fuel cell vehicle.** While electrification is key to mitigating the emissions from passenger and light freight vehicles, there is a limited scope for its use for heavy-duty commercial vehicles and shipping. Hydrogen-based fuel cell vehicles can be a potential solution for heavy-duty long distance freight transportation. Currently limited hydrogen distribution infrastructure limits the growth of HFC vehicles at the moment, but as more hydrogen production processes become commercially available, growth in the HFC vehicle segment is expected in medium term.

1 K. Yadav and A. Sircar. Hydrogen Compressed Natural Gas and Liquefied Compressed Natural Gas: Fuels for Future. *PDPU Journal of Energy & Management.*

2 World Economic Forum (WEF), 2023. Net-Zero Industry Tracker 2023 Edition Insight report, available online at https://www3.weforum.org/docs/WEF_Net_Zero_Tracker_2023_REPORT.pdf.

(vi) **Hydrogen-based energy storage.** Grid balancing with a high share of variable renewable energy in the grid is emerging as an issue. To overcome the grid stability issue, ensure maximum utilization of the renewable energy, and avoid curtailment of renewable energy-based power generation, electrolysis-based hydrogen production is a possible solution. Here, the surplus electricity from the renewable power sources is used to produce hydrogen to avoid curtailment and store energy.

These future potential developments may emerge as important over the period of the model, but are not included in the low-carbon scenario, as commercial applicability in the context of Pakistan up to 2050 is uncertain.

REFERENCES

ADB. 2019. *Energy Security in Pakistan: A Quantitative Approach to a Sustainable Energy Policy*. Manila. https://www.adb.org/sites/defaut/files/publications/535821/adbi-wp1024.pdf.

International Energy Agency (IEA). 2021. *Secure energy transitions in the power sector*. https://www.iea.org/reports/secure-energy-transitions-in-the-power-sector.

———. Pakistan. 2021. *IEA, Pakistan*. https://www.iea.org/countries/Pakistan.

National Energy Efficiency and Conservation Authority. 2020. *Strategic Plan (2020-2025)*. https://neeca.gov.pk/SiteImage/Misc/files/NEECA%20Strategic%20Plan%202020-25%20-Draft.pdf.

National Transmission and Despatch Company. 2022. *Indicative Generation Capacity Expansion Plan (IGCEP) 2022-31*. Prod. National Transmission and Despatch Company (NTDC).

Riahi, K., Van Vuuren, D.P., Kriegler, E., Edmonds, J., O'neill, B.C., Fujimori, S., Bauer, N., Calvin, K., Dellink, R., Fricko, O. and Lutz, W. 2017. The Shared Socioeconomic Pathways and their energy, land use, and greenhouse gas emissions implications: An overview. *Global Environmental Change*.

Siyal, A.W. and Gerbens-Leenes, P.W. 2022. The water–energy nexus in irrigated agriculture in South Asia: Critical hotspots of irrigation water use, related energy application, and greenhouse gas emissions for wheat, rice, sugarcane, and cotton in Pakistan. *Frontiers in Water* 4, p. 941722.

United Nations Framework Convention on Climate Change. 2015. *Pakistan's Intended Nationally Determined Contribution* (PAK-INDC). https://www4.unfcc.int/sites/submissions/INDC/Published%20Documents/Pakistan/1/Pak-INDC.pdf.

———. 2021. *Updated Nationally Determined Contributions 2021 - Pakistan*. https://unfccc.int/sites/default/files/NDC/2022-06/Pakistan%20Updated%20NDC%202021.pdf.